Human Biology and

T0228532

Published Symposia of the Society for the Study of Human Biology

Numbers 1–9 were published by Pergamon Press, Headington Hill Hall, Headington, Oxford OX3 0BY. Numbers 10–24 were published by Taylor & Francis Ltd, 10–14 Macklin Street, London WC2B 5NF. Numbers 25–40 were published by Cambridge University Press, The Pitt Building, Trumpington Street, Cambridge CB2 1RP. Further details and prices of back-list numbers are available from the Secretary of the Society for the Study of Human Biology.

Society for the Study of Human
Biology Series: 42

Human Biology and History

Edited by

Malcolm Smith

Department of Anthropology
University of Durham

CRC Press
Taylor & Francis Group
Boca Raton London New York

CRC Press is an imprint of the
Taylor & Francis Group, an **informa** business

A TAYLOR & FRANCIS BOOK

CRC Press
Taylor & Francis Group
6000 Broken Sound Parkway NW, Suite 300
Boca Raton, FL 33487-2742

First issued in paperback 2020

ISBN-13: 978-0-367-45486-9 (pbk)
ISBN-13: 978-0-415-28861-3 (hbk)

Visit the Taylor & Francis Web site at
http://www.taylorandfrancis.com

and the CRC Press Web site at
http://www.crcpress.com

Typeset in 10/12 pt Baskerville by
Graphicraft Limited, Hong Kong

British Library Cataloguing in Publication Data
A catalogue record for this book is available from the British Library

Library of Congress Cataloging in Publication Data
Human biology and history / edited by Malcolm Smith.
      p. cm. — (Society for the Study of Human Biology symposium series ; 42)
    Includes bibliographical references and index.
    ISBN 0-415-28861-4 (cloth : alk. paper)
    1. Physical anthropology and history.   2. Human remains (Archaeology).
  3. History—Methodology. I. Smith, Malcolm T. II. Series.
  GN62 .H86 2002
  599.9—dc21   2002020491

Contents

Figures

Tables

Contributors

Dr Pia Bennike
University of Copenhagen
Panum Institut
Laboratory of Biological Anthropology
Blegdamsvej 3
2200 Copenhagen Denmark

Professor Laura Betzig
Museum of Zoology
University of Michigan
Ann Arbor MI 48109
USA

Professor Roderick Floud
Provost, London Guildhall
University, 31 Jewry Street,
London EC3N 2EY,
UK

Dr Christopher J. Knüsel
Calvin Wells Laboratory
Department of Archaeological Sciences
University of Bradford
Bradford
West Yorkshire BD7 1DP
UK

Dr John Landers
All Souls College
Oxford OX1 4AL
UK

Professor James H. Mielke
Department of Anthropology
University of Kansas
Lawrence KS 66045
USA

Derek J. Oddy
University of Westminster
School of Social and Policy Sciences
Faculty of Business
Management and Social Studies
309 Regent Street
London W1R 8AL
UK

Dr Holger Schutkowski
Department of Archaeological Sciences
University of Bradford
Bradford
West Yorkshire BD7 1DP
UK

Dr Malcolm T. Smith
Department of Anthropology
University of Durham
43 Old Elvet
Durham DH1 3HN
UK

Preface

This volume has its origin in contributions to the symposium *Human Biology and History*, held at University College, Durham under the auspices of the Society for the Study of Human Biology. Researchers from a broad range of disciplines – from archaeology, anthropology, geography, biology and history – presented their work to an audience of equally mixed provenance, and stimulated and contributed to a wide-ranging discussion of methods for, theoretical approaches to and outcomes of research into the biological history of past human populations. The themes explored in this volume – the survival and reproduction of individuals and populations, the evidence of environment and nutrition on growth and development, the structure and mobility of populations of the past – emerged as areas of shared interest, in spite of being researched and evidenced by means of very different approaches. The achievement so far is heartening, even inspiring, but the studies reported in this volume are important for revealing the potential for interaction between human biology and history rather than setting limits to what can be achieved – we have secured the common ground, not yet the promised land.

My thanks are due to the authors and publishers, to the conference delegates, and to colleagues in the Department of Anthropology at Durham. University College, Durham is also Durham Castle, part of the World Heritage Site on the peninsula which also contains Durham Cathedral. One could scarcely imagine a situation better suited to stir the historical imagination, and the kindness and efficiency of the staff at Durham Castle helped us to make the most of our surroundings.

Financial support was provided by the Society for the Study of Human Biology, the University of Durham, and Glaxo-Wellcome, and is gratefully acknowledged.

Malcolm Smith

1 Human biology and history

The scope and scale of interaction

Malcolm Smith

Introduction

The purpose of this volume is to exemplify the range of common interest and to illustrate the scope and potential for interaction between the fields of human biology and history. The chapters in this collection reflect aspects of research current and established within both disciplines which address these common interests. The novelty of this compilation lies in its recognition that when the range of interdisciplinary contributions are assembled in this way, we can acknowledge the existence of a corpus of work which can be designated 'biological history'.

How do human biology and history interact? At the simplest level, human biologists can exploit historical data, that is, the written records of communities that lived in the past, in order to explore an agenda of significance to biology. Further, awareness of arguments drawn from historical debate can point to issues and circumstances that might have important influence on biological variables. At the biological level, there is often little conceptual difference between studies on historical populations and studies on contemporary populations. Thus, for every study of population genetics, or sociobiology, or growth or dietary energetics in the historical population, there will be equivalent studies on people living in the present day.

Beyond this, however, there is a value to historical records which makes them much more than simply a data source additional to observations made of the living population. First, of course, there is the intrinsic interest in historical material for the insight it can give into the lives of individuals and societies of the past. For many biological anthropologists this engagement with the people of the past is the motivation for study, and spurs the search for and refinement of techniques by which the past can be understood. It can stimulate the study of, say, stature, diet and disease in past populations. For some areas of human biology, on the other hand, there are practical advantages to historical data which give them a value not to be found in contemporary surveys, and motivate their study as the best data for investigating topics of biological interest. Because humans are such a long-lived species, we need extended runs of historical data to study secular trend, or to analyse topics such as fertility, mortality and inbreeding over a number of generations. Historical archives also permit us the opportunity to focus on specific periods when events with biologically significant consequences took place. There is also an ethical dimension to the difference between studying living populations and those from the past whose records have survived. While many who work with historical archives will be touched by the common humanity we share with our subjects, archival research is necessarily less intrusive, and less burdened with the responsibility to do well by the people we are studying.

The ways in which biology can contribute towards historical understanding and research may broadly be characterised thus: first, by offering techniques and methods, a new set of practical tools and approaches for addressing existing historical questions, and second, more controversially, by offering hypotheses derived from biology to interpret existing historical data. In terms of the material to be analysed, we can distinguish between two kinds of data sources: the written records accumulated by literate societies, and the various kinds of human remains, ranging in scale from mass burials to molecules, which can be scrutinised to describe in life the people they represent. This latter category has been, of course, the exclusive province of archaeologists and biological anthropologists, but may be seen as increasing relevance to historians who may call upon these analytical techniques to answer their own questions.

I have not included in this volume the relationship that links the disciplines through the history of science and history of biology. This is an important area of scholarship in biology, history and philosophy, perhaps nowhere more engaging to the interests of human biologists than in the study of the development of such pioneers of evolutionary biology and genetics as Darwin (Bowler 1990; Mayr 1991), Mendel (Henig 2000; Vitezslav 1996) and Galton (Keynes 1993), but it is not grist to our mill.

Contents

The chapters are ordered broadly to reflect the data sources, either records or remains, as described above. The first five authors – Landers, Betzig, Mielke, Smith and Oddy – base their analyses on the historical and biological interpretation of textual material. Floud's chapter spans the two, describing a corpus of work which exploits historical records of body measurements. It is thus located both within the world of historical records, and in the world of direct inference from human subjects. The remaining three chapters – by Bennike, Knüsel and Schutkowski – use a variety of methods, from the observation of gross anatomy to molecular analysis, to draw inferences by direct analysis of human remains from the archaeological record.

The written sources which can be interpreted to yield information about the biology of past populations are of various kinds. In England, for example, vital records, such as ecclesiastical records of births, marriages and burials, can be interpreted as surrogates for records of births, mate choice and deaths. Civil registration, too, provides a record of similar vital events. Techniques of aggregative and record-linking analysis enable the estimation of secular trend of mortality and fertility, and both these have been the focus of attention for social and economic historians as well as human biologists. Migration can be estimated from the records of lifetime mobility, inferred from censuses, marriage records and other archives like apprenticeship records, church court depositions and Poor Law accounts.

The first two chapters in this volume exemplify how an analytical approach grounded in evolutionary theory can both enrich and challenge interpretations of the demography of past populations. In the opening chapter, Landers addresses the issue of alternative paradigms for the interpretation of English demographic history. This comparative perspective is rare and valuable, since most researchers are active within either within the traditional framework of historical demography, derived from the Malthusian influence on economic history and social history, or from a biological perspective, which takes a much more ecological, epidemiological or evolutionary view of events. By discussing both Darwinian and Malthusian aspects of the concept of 'adaptation', Landers brings a fresh theoretical perspective to the analysis of the secular trend of vital events in England, which has generally been explained by the influence of such economic markers as prices and real wages on the trends in nuptiality, fertility and mortality.

The importance of biological theory as the basis for an approach to the history of marriage, mating and fertility in Britain is further developed in Betzig's chapter, which analyses very 'traditional' material – the lives and loves of rich and famous men – within the framework of Darwinian evolution. Betzig's work has its origins within the tradition of sociobiology, that is the attempt to explain social behaviour in terms of an evolutionary framework, and she has called this approach 'Darwinian History' (Betzig 1992). Though originally formulated in relation to organisms other than humans (Wilson 1975), the implications of sociobiology for the study of human beings were early appreciated (Chagnon and Irons 1979). Such work tests hypotheses based on the premise that mate choice, reproduction, parenting and the investment of resources in offspring should all be expected to maximise the reproductive success of the actors, and in this sense only such behaviour is referred to as 'adaptive'. 'Adaptation' here signifies not that some behaviour is genetically grounded, but that – by whatever proximate agency motivated – it has as its ultimate and perhaps unconscious goal the maximisation of reproductive success (Caro and Borgerhoff Mulder 1987), or of 'Darwinian fitness', as it is also termed. Evolutionary explanations based on this premise, have been offered for differential female infanticide in historical India and China (Dickemann 1979) and for the religious claustration of women in medieval Europe (Boone 1986).

The pattern of investment of resources in the next generation has also been analysed in this evolutionary framework (Judge and Hrdy 1992) and the same authors have scrutinised the inheritance pattern of primogeniture through the same lens (Hrdy and Judge 1993). Strategies of parental investment in relation to wealth and social status have also been examined in Europe (Voland *et al.* 1997). Betzig's style of research and writing (e.g. Betzig 1995) is somewhat different from the traditional methods of historical demography, characterised as it is by collation and interpretation of literary sources rather than vital records, and in this respect it seems closer to the studies making inference from literate elites, such as those of Boswell (1991) or, in particular, Stone (1977), whom she acknowledges as an influence.

In parallel with the development of historical demography by social and economic historians has been the exploitation of similar data sources by biological anthropologists with an interest in the genetic structure and genetic evolution of populations. Influenced by the French historical demographers and by the development of English historical demography at the Cambridge Group, biological anthropologists in Britain and the United States soon began to analyse data from historical vital records using models derived from population genetics theory. Such studies use correlates of migration and mate choice to predict the distribution of genes and genotypes in populations, and may also compare these outcomes to genetic data. Classic applications of these methods include studies of the populations of the Otmoor region of Oxfordshire, UK (Küchemann *et al.* 1967; Harrison 1995), the historical settlement of the Lower Connecticut Valley, Massachusetts (Swedlund *et al.* 1984, 1985), the Utah Mormons (Jorde 1989; O'Brien *et al.* 1994) and the Åland Islands (Mielke 1980; Mielke *et al.* 1994). Mielke's chapter in this volume carries forward the analysis of the Åland archipelago, with a discussion of the influence of the War of Finland (1808–9) on vital events and population size. Mielke shows that the excessive mortality on Åland during this period was caused by epidemic disease introduced by Swedish troops defending the Ålanders against the invading Russian forces, and concludes by discussing the genetic consequences of these and other demographic events for the genetics of the Åland population.

Smith's research, too, is based in the practice of biological anthropology derived from population genetics. The study of inbreeding has long been a focus of biological anthropology, since it can have a substantial impact on the distribution of genes and genotypes, and consequent effects on the pattern of genetic disease. In the historical populations of Europe and its colonies, the existence of records of dispensations allowing members of the Roman Catholic Church to marry within the prohibited degrees of consanguinity furnished human biologists with a unique archive for the study of inbreeding over a period of several centuries. Beyond the influence of the Roman Catholic Church such data are rarely available, and inbreeding is difficult to estimate, requiring the compilation of genealogies from vital records. Crow and Mange (1965) developed a method for estimating inbreeding from the frequency of occurrence of marriages between persons bearing the same surname. Smith's chapter describes the development of further methods used extensively in biological anthropology for the analysis of population genetic structure based on the distribution of surnames, and suggests that these methods might find application to answer more broadly based questions formulated within historical debate.

Oddy's chapter addresses how the nutritional status of past populations might be estimated from historical sources. This topic occupies a pivotal position between social and cultural history, physiology and the study of anthropometrics. A great deal has been written about historical diet, including the chronological and seasonal availability of foodstuffs and the social and economic differentiation of diet, but it falls far short of quantitative measurements necessary to estimate energy intake with accuracy. Research on dietary energetics constitutes an important area of human biology of living populations (Ulijaszek 1995, 1996), as do studies of energy expenditure in relation to specific physical activities (Elton *et al.* 1998), ecology and occupation (Leonard *et al.* 1996) and life cycle (Rashid and Ulijaszek 1999). Accurate measures of dietary intake are notoriously hard to attain (Huss-Ashmore 1996), and the ascertainment of the energetics of historical diet is especially difficult, because of the lack of available data. There are problems of two sorts here, one is the scarcity of reliable source material, and the other is the translation of historically described foodstuffs in compositional terms that are equivalent to their modern, measurable counterparts. One kind of source that has been exploited relates to the bookkeeping associated with the diets of institutions such as prisons and workhouses; these are of interest because they not only specify the items of diet but also prescribe quantities of different foodstuffs. They are, however, not necessarily representative of diets among those outside such institutions; indeed, as Crawford (1993) has pointed out, workhouse diets were deliberately designed to be less nutritious than those of people not eligible for such relief. In this volume, Oddy analyses the much scarcer and potentially more revealing data which came from rare surveys of family budgets, and through analysis of the energetics of these data, he infers not only regional differences in dietary habits, but also trends in dietary composition from the eighteenth to the twentieth centuries.

The significance of detailed dietary studies can readily be appreciated in the context of the chapter which follows, Floud's overview of the development of the subject of anthropometric history. This term has been given to the practice of analysing historical series of body measurements in the light of the economic and social circumstances of the subjects (Floud *et al.* 1990). One foundation of this young and growing research area was the pioneering research in human biology of Tanner and his associates on human growth and development, including secular trend in stature (Tanner 1981, 1982). Using Tanner's work as a baseline, economic historians were able to interpret records of stature of individuals recruited to the armies, prisons and other institutions to address the agenda of their own concerns regarding the effects of wages and socio-economic factors on achieved height. Research on anthropometric history has been extended across the continents of Europe, Asia, Australia and North America (Komlos and Baten 1998), and into specific historical contexts, such as that of Ireland in the decades surrounding the Great Famine of 1846–51 (Mokyr and Ó Gráda 1996; Jordan 1998), or the United States of America before the civil war (Komlos and Caclanis 1997).

The remaining chapters are concerned with inference through the direct analysis of skeletal remains. While these authors' methods differ markedly from the analytical techniques used with historical records, and address a broad range of issues, there is a congruence of research goals with the research described above.

Bennike's work is firmly grounded in the tradition of palaeopathology, the elucidation of health status, including development, disease and trauma, from skeletal evidence. Such evidence may indicate the presence in past populations of a wide range of physical conditions, including traumas and degenerative and infectious diseases (Roberts and Manchester 1995). While it has been recognised that caution is necessary in making inferences about the health status of populations (Wood *et al.* 1992) such evidence allows not only individual diagnoses but also inference about epidemiology and process, for example global spread diseases.

In her comparison of pairs of populations ostensibly sharing the same environment Bennike also addresses the debate about the causes of 'success' or 'failure' of past communities as a result of evolutionary or cultural adaptation. To what extent can a detailed account of culture or phylogeny explain the demise or survival of populations? In discussing this issue Bennike reaches further back into prehistory than any other contribution to explain the fate of the Neanderthals, people of the European Mesolithic, and the Greenland Norse.

Another category of skeletal memory records not disease, but rather the plastic responses of the skeleton to habitual activity. These changes, including characteristic developments produced by occupational stresses (Jurmain 1999), have been recognised in a number of habitual postures, such as the kneeling in prayer of monks (Sheridan 1997), and squatting (Boulle 2001). As with the palaeopathology of disease, interpretations of such plasticity data can go beyond individual inference towards an understanding of process, as for example in Bouchet's use of the chronology and frequency of squatting facets to infer the availability of domestic furniture for sitting on (Bouchet 1989). Knüsel's chapter reviews the history of such observations of the effects of occupation on the skeleton from the late eighteenth century to the present, reflecting on the recent loss of many trades which left their imprint on the skeleton, and the largely recent development of bilateral assymetries associated with elite sports performance.

The molecular study of human remains is a new and fast-growing field, which enables exploration of the identities, relationships, environments, diets and behaviours of individuals and groups in the past. With increasing success, aspects of DNA are tractable to analysis from samples derived by historical and ancient sources (Herrmann and Hummel 1994). This means that DNA can be used to test individual identifications from remains, as in the case of Joseph Mengele (Jeffreys *et al.* 1992), or the assassinated Romanov family (Gill *et al.* 1994), or Louis XII of France (Jehaes *et al.* 2001). Collaborations between historians and human geneticists have also resulted in an exhaustive analysis of claims that the porphyria gene runs in the European Royal families, though in the published account the documentary evidence is perhaps more compelling than the genetic (Röhl *et al.* 1998).

The analysis of chemical elements taken up into the body from the environment can yield evidence about dietary components, and by extension, behaviour, of individuals. For example, trace elements such as lead or strontium can be used to reconstruct dietary components and to reveal characteristics of the environment. Since tooth enamel is laid down in childhood, trace elements in dental tissue can be used to estimate lifetime migration (Montgomery *et al.* 2000). The protein collagen, extracted from bone, can be tested for the isotopes of carbon

(^{12}C and ^{13}C) and of nitrogen (^{15}N and ^{14}N). The isotope ratios can reveal dietary components; for example, the proportion of animal versus plant protein can be inferred from the nitrogen isotope ratio (δ^{15}N), and the relative importance of marine or terrestrial products in the diet is reflected in the carbon isotope ratio (δ^{13}C). As with other techniques described here, these basic data can be interpreted in ways which extend their application to reveal aspects of social structure and process. By comparing mineral analyses of individual skeletons with grave goods (Schutkowski and Hermann 1999), inferences may be made about association of diet with social status or gender, as has been postulated in the late Roman period burials at Poundbury (Richards and Hedges 1998).

Schutkowski's contribution to this volume analyses medieval burials from Germany to exemplify a range of inferences from trace elements and isotopes, including the analysis of bone to investigate heavy metal burden associated with lead-smelting, the interpretation of strontium levels to measure lifetime migration, and the deduction of dietary components from carbon and nitrogen.

Conclusion

This kind of 'biological history' is not a homogeneous discipline, with a single method or theoretical orientation, but, as will become plain, is almost as diverse as the spheres from which it is derived. Almost, but not quite, for this is understandably not an enterprise to suit all historians any more than it will appeal to all human biologists. The contribution to history which I see biology making is firmly within a positivist framework, a world where there are ascertainable facts, and testable hypotheses. We could draw the analogy with anthropology and archaeology, since both disciplines have branches reliant on biological science as well as areas grounded in social theory. It is not surprising, therefore, that the branches of history which have most thoroughly embraced biology, to the extent that the historians are directly involved in the analyses, are economic and demographic history. In spite of the perennial appeal of the challenge, the work described here can make little claim to bridge in any profound sense C. P. Snow's two cultures (Snow 1959, 1993), but I hope that it may both reflect and further stimulate the traffic of communication between scholars and researchers who, though raised in different traditions, find themselves addressing common problems and sharing common aspirations in the study of our forebears as biological and social beings.

References

Betzig, L. (ed.) (1992) *Darwinian History*, special issue of *Ethology and Sociobiology*, 13.
Betzig, L. (1995) 'Medieval monogamy', *Journal of Family History*, 20, 181–216.
Boone, J. L. (1986) 'Parental investment and elite family structure in preindustrial states: a case study of late Medieval–Early Modern Portuguese genealogies', *American Anthropologist*, 88, 859–78.
Boswell, J. (1991) *The Kindness of Strangers*, New York: Pantheon.

Bouchet, L. (1989) 'L'usage de la station accroupie dans les sociétés antiques et médiévales de Gaule', *Actes 4èmes Journées Anthropologiques*, Paris: Centre National de la Recherche Scientifique, pp. 114–22.

Boulle, E-L. (2001) 'Evolution of two human skeletal markers for the squatting position: a diachronic study from antiquity to the modern age', *American Journal of Physical Anthropology*, 115, 50–6.

Bowler, P. J. (1989) *Evolution: the history of an idea*, Berkeley: University of California Press.

Bowler, P. J. (1990) *Charles Darwin: the man and his influence*, Oxford: Blackwell.

Caro, T. and Borgerhoff Mulder, M. (1987) 'The problem of adaptation in the study of human behaviour', *Ethology and Sociobiology*, 8, 61–72.

Chagnon, N. and Irons, W. (eds) (1979) *Evolutionary Biology and Human Social Behaviour*, North Scituate, MA: Duxbury.

Crawford, E. M. (1993) 'The Irish workhouse diet, 1840–90', in Geissler, C. and Oddy, D. J. (eds) *Food, Diet and Economic Change Past and Present*, Leicester: Leicester University Press, pp. 83–100.

Crow, J. F. and Mange, A. P. (1965) 'Measurement of inbreeding from the frequency of marriages of persons of the same surname', *Eugenics Quarterly*, 12, 199–203.

Dickemann, M. (1979) 'Female infanticide, reproductive strategies, and social stratification: a preliminary model', in Chagnon, N. and Irons, W. (eds) *Evolutionary Biology and Human Social Behaviour*, North Scituate, MA: Duxbury, pp. 312–67.

Edelson, E. (1999) *Gregor Mendel, and the Roots of Genetics*, Oxford: Oxford University Press.

Elton, S., Foley, R. and Ulijaszek, S. J. (1998) 'Habitual energy expenditure of human climbing and clambering', *Annals of Human Biology*, 25, 523–31.

Floud, R., Wachter, K. W. and Gregory, A. (1990) *Height, Health and History: nutritional status in the United Kingdom, 1750–1980*, Cambridge: Cambridge University Press.

Gill, P., Ivanov, P. L., Kimpton, C., Piercy, R. *et al.* (1994) 'Identification of the remains of the Romanov family by DNA analysis', *Nature Genetics*, 6: 130–5.

Harrison, G. A. (1995) *The Human Biology of the English Village*, Oxford: Oxford University Press.

Henig, R. M. (2000) *A Monk and Two Peas: the story of Gregor Mendel and the discovery of genetics*, London: Weidenfeld & Nicolson.

Hermann, B. and Hummel, S. (1994) *Ancient DNA: recovery and analysis of genetic material from paleontological, archaeological, museum, medical, and forensic specimens*, Heidelberg: Springer-Verlag.

Hrdy, S. B. and Judge, D. S. (1993) 'Darwin and the puzzle of primogeniture', *Human Nature*, 4, 1–45.

Huss-Ashmore, R. (1996) 'Issues in the measurement of energy intake for free-living human populations', *American Journal of Human Biology*, 8, 159–67.

Jeffreys, A. J., Allen, M. J., Hagelberg, E. and Sonnberg, A. (1992) 'Identification of the skeletal remains of Mengele, Josef by DNA analysis', *Forensic Science International* 56: 65–76.

Jehacs, E., Pfeiffer, H., Toprak, K., Decorte, R., Brinkmann, B. and Cassiman, J. J. (2001) 'Mitochondrial DNA analysis of the putative heart of Louis XVII, son of Louis XVI and Marie-Antoinette', *European Journal of Human Genetics*, 9, 185–90.

Jordan, T. E. (1998) *Ireland's Children: quality of life, stress, and child development in the famine era*, Westport, CT: Greenwood Press.

Jorde, L. B. (1989) 'Inbreeding in the Utah Mormons: an evaluation of estimates based on pedigrees, isonymy and migration matrices', *Annals of Human Genetics*, 53, 339–55.

Judge, D. S. and Hrdy, S. B. (1992) 'Allocation of accumulated resources among close kin: inheritance in Sacramento, California, 1890–1984', *Ethology and Sociobiology*, 13, 495–522.

Jurmain, R. (1999) *Stories from the Skeleton: behavioral reconstruction in human osteology*, London: Taylor & Francis.

Keynes, M. (ed.) (1993) *Sir Francis Galton, FRS*, London: Macmillan.

Komlos, J. and Baten, J. (eds) (1998) *The Biological Standard of Living in Comparative Perspective*, Stuttgart: Franz Steiner Verlag.

Komlos, J. and Caclanis, P. (1997) 'On the puzzling cycle in the biological standard of living: the case of antebellum Georgia', *Explorations in Economic History*, 34, 433–59.

Küchemann, C. F., Boyce, A. J. and Harrison, G. A. (1967) 'A demographic and genetic study of a group of Oxfordshire villages', *Human Biology*, 39, 251–76.

Leonard, W. R., Katzmarzyk, P. T. and Crawford, M. H. (1996) 'Energetics and population ecology of Siberian herders', *American Journal of Human Biology*, 8, 275–89.

Mayr, E. (1991) *One Long Argument*, Cambridge, MA: Harvard University Press.

Mielke, J. H. (1980) 'Demographic aspects of population structure in Åland', in Eriksson, A. W., Forsius, H. R., Nevalinna, H. R., Workman, P. L. and Norio, R. (eds) *Population Structure and Genetic Disorders*, London: Academic Press, pp. 471–86.

Mielke, J. H., Relethford, J. H. and Eriksson, A. W. (1994) 'Temporal trends in migration in the Åland Islands – effects of population-size and geographic distance', *Human Biology*, 66, 399–410.

Mokyr, J. and Ó Gráda, C. (1996) 'Height and health in the United Kingdom 1815–1860: evidence from the East India Company Army', *Explorations in Economic History*, 33, 141–68.

Montgomery, J., Budd, P. and Evans, J. (2000) 'Reconstructing the lifetime movements of ancient people: a Neolithic case study from southern England', *European Journal of Archaeology*, 3, 407–22.

O'Brien, E., Rogers, A. R., Beesley, J. and Jorde, L. B. (1994) 'Genetic-structure of the Utah Mormons – comparison of results based on RFLPs, blood-groups, migration matrices, isonymy, and pedigrees', *Human Biology*, 66, 743–59.

Rashid, M. and Ulijaszek, S. J. (1999) 'Daily energy expenditure across the course of lactation among urban Bangladeshi women', *American Journal of Physical Anthropology*, 110, 457–65.

Richards, M. P. and Hedges, R. E. M. (1998) 'Stable isotope analysis reveals variations in human diet at the Poundbury Camp cemetery site', *Journal of Archaeological Science*, 25, 1247–52.

Roberts, C. and Manchester, K. (1995) *The Archaeology of Disease*, 2nd edn, Ithaca, NY: Cornell University Press.

Röhl, J. C. G., Warren, M. and Hunt, D. (1998) *Purple Secret*, London: Bantam Press.

Schutkowski, H. and Hermann, B. (1999) 'Diet, status and decomposition at Weingarten: trace element and isotope analyses on early medieval skeletal material', *Journal of Archaeological Science*, 26, 675–85.

Sheridan, S. (1997) 'Biocultural reconstruction of kneeling pathology in a Byzantine Judean monastery', *American Journal of Physical Anthropology*, supplement 24, 209.

Snow, C. P. (1959) Rede Lecture, University of Cambridge.

Snow, C. P. (1993) *The Two Cultures*, Cambridge: Cambridge University Press.

Stone, L. (1977) *The Family, Sex and Marriage in England 1500–1800*, London: Weidenfeld & Nicolson.

Swedlund, A. C., Jorde, L. B. and Mielke, J. H. (1984) 'Population structure in the Connecticut Valley. I. Marital migration', *American Journal of Physical Anthropology*, 65, 61–70.

Swedlund, A. C., Anderson, A. B. and Boyce, A. J. (1985) 'Population-structure in the Connecticut Valley. 2. A comparison of multidimensional-scaling solutions of migration matrices and isonymy', *American Journal of Physical Anthropology*, 68, 539–47.

Tanner, J. M. (1981) *A History of the Study of Human Growth*, Cambridge: Cambridge University Press.

Tanner, J. M. (1982) 'The potential of auxological data for measuring economic and social well-being', *Social Science History*, 6, 571–81.

Ulijaszek, S. J. (1995) *Human Energetics in Biological Anthropology*, Cambridge: Cambridge University Press.

Ulijaszek, S. J. (1996) 'Energetics, adaptation and adaptability', *American Journal of Human Biology*, 8, 169–82.

Vitezslav, O. (1996) *Gregor Mendel: the first geneticist*, Oxford: Oxford University Press.

Voland, E., Dunbar, R. I. M., Engel, C. and Stephan, P. (1997) 'Population increase and sex-biased parental investment in humans: evidence from 18th- and 19th-century Germany', *Current Anthropology*, 38, 129–35.

Wilson, E. O. (1975) *Sociobiology: the new synthesis*, Cambridge, MA: Belknap Press.

Wood, J. W., Milner, G. R., Harpending, H. C. and Weiss, K. M. (1992) 'The osteological paradox: problems of inferring prehistoric health from skeletal samples', *Current Anthropology*, 33, 343–70.

2   Adaptation and the English demographic regime

John Landers

Introduction

Darwinian theory relies heavily on population concepts. Populations are the units of evolution, and neo-Darwinism defines the evolutionary process in terms of changes in their genetic composition. Conventional accounts of Darwinian origins see its founder's exposure to the 'dismal science' of the classical economists – above all Malthus with his demographic arguments – as the inspiration for the notions of natural selection and the 'struggle for life'. But important as these are, they form only one of two conceptual bases of Darwinian theory. Darwin is said to have derived the second, the concept of adaptation, from the utilitarian natural theology of Paley which had its intellectual roots in the same soil as those of classical economics and thus of Malthus himself (Bowler 1989). Certainly the notion of adaptation – if not under that explicit title – plays an important part in Malthusian theory. There is an irony here. Malthus is still widely seen as a prophet of doom – of inevitable mass poverty, if not of actual famine and starvation – but he was, by the standards of his contemporaries and immediate predecessors, a relative optimist (Wrigley 1983, 1987a). Particularly in his later work, he considered whether popular living standards could be protected, or even improved, given the limitations of food production and the consequences of the 'necessary passion between the sexes', and his response was guardedly positive. Providing marriage was delayed or forgone where economic circumstances were unpromising, the supply and demand for labour could be balanced against each other and the 'positive check' of rising mortality held at bay (Wrigley 1986a). In this way the demographic behaviour of populations could be adapted to the limitations of the economic space they occupied.

Adaptation and the organic economy

These 'Darwinian' and 'Malthusian' concepts of adaptation thus emerge from a common intellectual nexus, but in all except perhaps the longest term, their implications are quite distinct if not wholly opposed. Applying either of them in practice requires specifying three elements: a specific trait or complex of traits, a specific environment, and a criterion of adaptedness. To work backwards, our two distinct concepts of adaptation imply correspondingly distinct adaptive criteria. In the Darwinian case the issue is one of demographic expansion, and a trait is adaptive on this criterion if it enables its possessors to be fruitful and multiply more rapidly than they would otherwise be able to and to sustain this differential growth over an extended period. By contrast the Malthusian criterion is economic and refers to a population's level of per capita output and thus to its standard of living. Maximising this quantity generally requires the restriction of population growth, at least in the medium term. As well as specifying the criteria of adaptation we also have to consider the level at which it operates. In other words, does the adaptation benefit individuals themselves or the population as a whole?

In the Malthusian case the issue is largely an empirical one of distribution and concerns the degree to which higher average incomes are translated into benefits for the generality of the population rather than being appropriated by the few. For Darwinian adaptation, however, this problem of 'macro-' versus 'micro-' levels raises a theoretical question which has catalysed the recent development of evolutionary theory. In its original form Darwinian theory was compatible with selection either at the level of the individual or of the group, and where behaviour was concerned this ambiguity survived the formation of the 'neo-Darwinian' synthesis. Wynne-Edwards (1962), in particular, argued that some traits were 'altruistic', since they damaged individual reproductive fitness while benefiting that of the local population, and had arisen by a process of 'group selection' (Wynne-Edwards 1962). The 'second-wave' neo-Darwinism of the 1970s rejected this explanation, insisting that phenomena such as territorial or hierarchical reproductive systems could be explained in terms of selection at the level of the individual gene (Dawkins 1976). But both culture and genetics influence human behaviour and, as advocates of the 'meme' concept remind us, traits can survive and prosper without necessarily enhancing the fitness of individuals. Hence we need to consider both macro- and micro-levels of potential adaptation (Blackmore 1999).

The organic economy

The main feature of the environment to which pre-industrial populations had to adapt was the low level of productivity imposed by a reliance on organic sources for nearly all inputs of raw materials and, above all, of energy. The latter were mostly filled by muscle power, often human, with wood as the major heat source. The result was what the economic and demographic historian Sir Tony Wrigley has termed an 'organic economy', characterised by multiple interacting supply constraints.[1] Since raw materials and energy supplies ultimately derived, like food supplies, from plant production, they were all potential competitors for the available land. Land used for fuel wood, or industrial crops, was not available for staple foods. Domestic animals such as horses or oxen, the main source of non-human motive power, required fodder, which took up land otherwise available for food production. Wind and water power might provide a 'land-free' alternative, but the number of suitable locations for mill races was limited by relief and the needs of water transport, and steady strong winds were not always available.

Limited energy supplies restricted overall output, particularly in heat-intensive processing and, above all, in metalworking. Deprived of the wherewithal for large-scale production, shortages of metal tools were pervasive in pre-industrial economies, and workers were often forced back onto a technology of wood which further limited productivity. But however serious these supply-side limits may have been in principle, it is likely that, in practice, productivity growth was more often limited by inadequate demand. Initially low productivity rendered mass poverty endemic, and so the great bulk of consumption – and thus of expenditure in monetarised economies – was needed simply to keep body and soul together. This left very little money available for non-subsistence goods and services, and so there was little incentive to produce them or for the technological innovations that might have increased efficiency or, indeed, ensured the full utilisation of factors of production such as labour. This was particularly important because the seasonal pattern of agricultural work was often very 'peaky'. Both capital, particularly draft animals, and labour might thus be unproductive through the slack periods, and pre-industrial economies often displayed a paradoxical combination of underemployment and scarcity.[2]

Bleak though the prospects for pre-industrial productivity may appear when viewed statically, things get worse when we consider the response to population growth over time. As Malthus and his contemporaries were only too aware, agricultural production was 'inelastic' inasmuch as it was very difficult to increase output in order to match rising demand in the long term.[3] Classical economics formulated this problem in terms of diminishing marginal returns. At the 'extensive margin', agricultural production might be increased by expanding the area under the plough, but since the best land was likely to be used first, later additions would yield less and so returns would tend to fall. At the 'intensive margin', production could be raised through increases in labour supply, but limited stocks of land and capital mean that, beyond a certain point, the scope for productive employment would fall away and each additional worker would produce less than his predecessor. As Malthus recognised, if this process continued long enough the marginal product would eventually fall to subsistence levels.

At this point, in a commercialised agriculture, further workers would have to be paid more than they produced, so labour force growth would cease, and subsequent population growth would have to be absorbed in other sectors or go to swell the ranks of the unemployed. But pre-industrial agriculture was not usually fully commercial and the problem of population growth was experienced in a different way. Frequently, production was organised on a familial basis around parcels of productive assets such as farms or craft workshops. These 'hearths', as they are generically termed, were transmitted between generations according to formal or customary rules that varied between societies but commonly resulted in impartible inheritance. Under these circumstances, where hearths were passed on as undivided units,[4] their supply was sufficiently inelastic as to be considered fixed, and the population divided into two categories. The first, the stable 'core', occupied hearths in the traditional familial economy, while the second eked out a living on its margins – as wage labourers if they were fortunate, or as vagrants if they were not. In such a context even modest rates of population growth could lead to profound structural change. For instance, a population that grew by only a half of 1 per cent annually would expand by 13 per cent over 25 years, but if the 'marginal' category accounted for an initial 5 per cent, it would quadruple in absolute numbers. Over a century, total population would grow by little more than 60 per cent, but the marginal category would expand by a factor of 14 and account for nearly half of the final total. Prolonged population growth thus posed a challenge that was as much social and political, potentially, as it was economic, by creating, not just additional persons, but an essentially new kind of person, a new class whose members lay substantially outside the existing order and might pose a powerful challenge to its stability (Goldstone 1991).

Urbanisation represented one possible solution to the problem of the marginal population. It could also contribute to productivity growth by fostering regional and occupational specialisation. The first required cities as *entrepôts* and centres of finance and communications, while the second generally involved non-agricultural occupations of a kind that flourished in urban environments. But here again, the reliance on muscle power imposed constraints that were manifested in the cost of hauling low-value, high-bulk commodities, such as staple foods, over land. Essentially, draft animals had to eat, and so the price of a cartload of grain would double over the distance it took the team to consume an equivalent volume of fodder. High-value, low-bulk commodities might be transported very long distances economically, but staple foods could rarely be moved more than 50 miles or so.[5] Hence urban populations usually had to feed themselves from a very limited territory unless they had access to water transport or were provisioned by powerful governments.

Demographic regimes

If the adaptive environment with which we are concerned is set by the structure of an organic economy, the trait complex is constituted by what demographers term a 'demographic regime'. These represent sets of relationships between variables that remained relatively stable over the long term, even though the variables themselves might fluctuate substantially. Alternatively, they might be seen on an individual level as enduring propensities to respond to particular circumstances in particular ways (Landers 1993: Chapter 1). Demographers usually regard the key population variables as nuptiality (the propensity to marry), marital fertility and mortality, but in this context the importance of urbanisation is such that we need also to include migration. With rare exceptions, the pattern of age-specific marital fertility in Europe before the late nineteenth century shows that, if any form of birth control was being practised, it was used to space successive births rather than to terminate childbearing prematurely as in modern populations. Average birth intervals varied greatly between regions, but they were remarkably stable over time, and much, if not all, of the regional variation was due to the physiological effects of differences in methods of infant feeding (Landers 1994).

Extramarital fertility was rarely of any demographic significance, and so birth rate variations primarily reflected variations in female age at marriage, proportions marrying, or both.

As a fully elaborated concept the demographic regime emerged from historical demographic work in recent decades, but it rests on ideas developed by Malthus himself. For Malthus, as we have seen, the differential expansive powers of population and the means of subsistence posed the central material problem of human existence. Starting with a deductive approach based on geometric and arithmetical ratios of increase, he progressed to a more detailed inductive analysis of population growth and agricultural production. But throughout his work he remained preoccupied with the power of population growth, a power which must eventually be checked by a reduction in either life span ('the positive check') or the relative numbers of births (the 'preventive check'). These two checks, and the feedback relationships that they imply, underlie a basic distinction between two kinds of demographic regime.

At issue is the response of birth rates to the economic consequence of diminishing returns to agricultural labour. If they remain constant, then population will continue to grow, and living standards to fall, until death rates rise sufficiently to terminate population growth at a 'high-pressure' equilibrium with elevated vital rates and living standards close to the margin of subsistence. Evidently, the lower the initial birth rate, the lower the equilibrium death rate and the higher the corresponding standard of living, but the best results ensue where birth rates themselves fall as living standards are eroded. In such a 'low-pressure' regime growth may cease before death rates have risen very much, if at all, and the equilibrium living standard will be higher. Malthus made this point forcibly in his advocacy of 'moral restraint' from marriage, but modern demographic analysis has shown that low birth rates also reduce the proportion of dependent children. Since this is likely to be matched by a greater probability of survival to adult life, investment in 'human capital' – education and training – will be stimulated. By contrast, in a high-pressure regime, the relative scarcity of resources, and the reduced likelihood that investment will be repaid in adulthood increases the pressure to obtain productive work from children at an early age and correspondingly reduced the scope for investment. Hence the question of how adaptive England's demographic regime was in Malthusian terms resolves itself into the question of whether it was of the high- or low-pressure variety.

The English case: population and economy

We have a relatively detailed picture of England's demographic history in the final centuries of the pre-industrial era, the so-called 'early-modern period'. Although 'modern' demographic statistics date only from the 1840s, there is an unusually long run of ecclesiastical parish registers of burials, baptisms and marriages originating in the middle of the sixteenth century. These present problems of coverage and quality, but Wrigley and Schofield (1981) have overcome many of these and applied the method of aggregate back projection (ABP) to material from a sample of 404 parishes. The resulting estimates of population size, expectation of life at birth (E0) and gross reproduction ratios (GRR)[6] can be supplemented by material from a further 30 or so parishes obtained using the method of family reconstitution, a form of nominal record linkage which allows the calculation of marital fertility rates and infant and child mortality by age. More recently Wrigley *et al.* (1997) have constructed a revised set of demographic estimates for England by applying a refined version of ABP termed generalised inverse projection (GIP) to the reconstitution sample. It is with these that we shall chiefly be concerned.[7]

Population growth

England's population roughly quadrupled from the mid-sixteenth to the early nineteenth centuries. Two waves of growth were separated by a century or so of stagnation (see Table 2.1) and so it is possible to compare the behaviour of wages and prices under very different demographic conditions. Initially, the results powerfully vindicate the expectation that prolonged population growth should depress living standards. Quinquennial population growth rates plotted against those in prices, as in Figure 2.1a, show a strongly positive relationship. Any increase in population is associated with increased prices until the close of the eighteenth century – ironically, the era of Malthus himself. But nominal price rises do not tell the whole story. Money wages also increased, and the plot of population growth against so-called 'real wages' in Figure 2.1b gives a rather different impression. The figure remains clearly oriented in the 'Malthusian' direction, but it now crosses the horizontal axis some way to the right of the origin, suggesting that labour demand expanded sufficiently to absorb the consequences of annual population growth of up to around a half of 1 per cent.

The two waves of population growth thus had different consequences. The first led to substantial pressure on living standards and was not sustained. The pressure is detectable in more detailed price series that show consumers redirecting expenditure from relative luxuries towards essential subsistence requirements as prices rose. Staple food thus increased in price by more than other items, and, within the basket of staples, the less favoured grains such as oats and barley rose by more than did wheat, the staple of choice (Clay 1984: 49–52).

Table 2.1 English population estimates from GIP

Year	Population (million)	Period	Growth (% p.a.)
1541	2.83	1541–1651	0.57
1651	5.31	1651–1751	0.11
1751	5.92	1751–1821	0.95
1821	11.46		

Source: Wrigley *et al.* 1997: Table A9.1

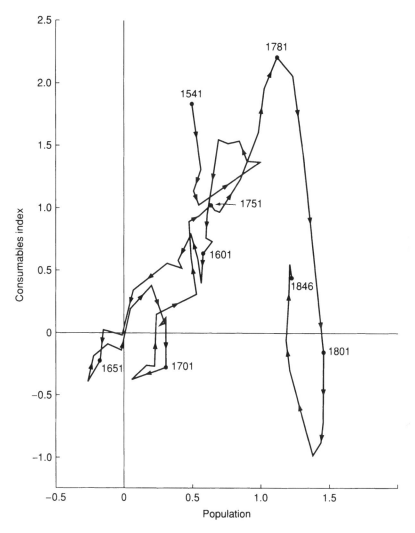

Figure 2.1a Rates of increase in English prices and population
Source: Wrigley and Schofield 1981: Table 10.2

Population and economy were brought back into balance by the cessation of population growth. Real wage levels then recovered until, by the second quarter of the eighteenth century, they exceeded those recorded at the beginning of the period (see Figure 2.2). By contrast, the second wave of population growth saw a response from the economy which enabled real wages to be maintained into the early nineteenth century. Furthermore, this cycle of population growth continued into the twentieth century, accompanied by radical economic and social transformation, and was terminated only by the spread of family limitation within marriage.

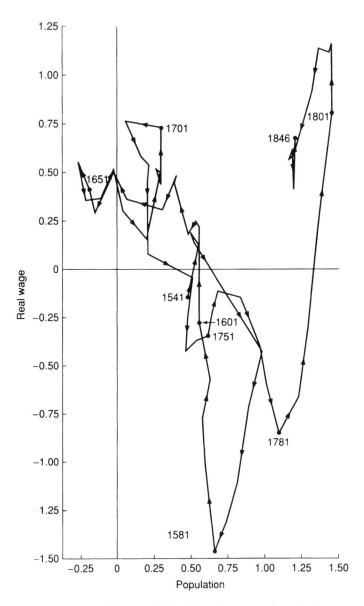

Figure 2.1b Rates of increase in English real wages and population
Source: Wrigley and Schofield 1981: Table 10.4

The first wave of population growth also affected the composition of the population in the expected manner. Detailed, or accurate, social structure data are absent for this period, but the work of Gregory King provides some contemporary estimates from a point close to the cessation of population growth. Writing at the end of the seventeenth century, King claimed that the families of

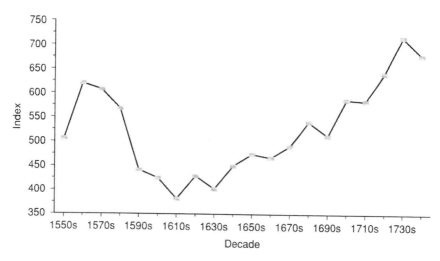

Figure 2.2 Real wages: decadal averages
Source: Wrigley and Schofield 1981: Appendix 9

paupers, labourers or 'cottagers' – those needing wages or welfare payments to survive – accounted for more than half of all families (Laslett 1973). It would be wrong to take King's figures at face value, but the fact that a well-informed contemporary expected such an assertion to be believed indicates the scale of change from the early sixteenth century when the overwhelming bulk of the population are likely to have been self-sufficient agricultural producers. The data on urbanisation given in Table 2.2 are similarly estimates, but again the contrast is clear. During the period of population growth, the urban proportion had expanded from around 5 per cent to 14 per cent. Most striking is the disproportionate growth of London, which continued for some decades into the era of stagnation, so that by 1700 the capital held about one-tenth of England's population and nearly two-thirds of its urban population.

Table 2.2 England's urban population (places >5,000)

	Absolute size (000s)			As % of national population		
Date	London	Other	Total	London	Other	Total
1520	55	70	125	2.25	3.00	5.25
1600	200	135	335	5.00	3.25	8.25
1670	475	205	680	9.50	4.00	13.50
1700	575	275	850	11.50	5.50	17.00
1750	675	540	1,215	11.50	9.50	21.00
1801	960	1,420	2,380	11.00	16.50	27.50

Source: Wrigley 1987c: Table 7.2

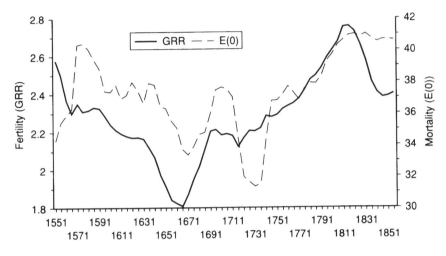

Figure 2.3 Fertility and mortality in England
Source: Wrigley *et al.* 1997: Table A9.1

Fertility and mortality

Fertility and mortality both varied substantially (see Figure 2.3). Life expectation fluctuated broadly in the range 30–40 years at birth and the GRR between roughly 1.8 and 2.8. There is no evidence for appreciable changes in age-specific marital fertility before the later eighteenth century, so we can assume that these movements were due to nuptiality. David Weir (1984) has shown that, by 'splicing' local family reconstitutions with ABP results, nuptiality change can be partitioned into components due to female age at marriage and proportions ever-married. The two cycles diverged in this respect. In the seventeenth century fertility decline stemmed from a fall in the proportion marrying. This ran at around 90 per cent for most of our period, but in the first half of the seventeenth century it fell sharply to trough at less than 80 per cent among female cohorts born in the 1610s (Weir 1984; Schofield 1989). In the late eighteenth century, by contrast, the increase in fertility was mainly due to younger marriage (see Figure 2.4). Overall, fertility and mortality made a roughly equal contribution to changes in the rate of population growth over the first cycle, but fertility played the predominant role after 1750.

Mortality and fertility both varied, but the issue that most concerns us is how they related to economic circumstances, and whether the 'high-' or 'low-pressure' models best describe England's demographic regime. On the nuptiality side, it is clear that Malthus's 'preventive check' operated during the first wave of growth since proportions marrying fell as economic circumstances deteriorated, and this made a major contribution to the cessation of growth. Similarly, on the mortality side, the pattern is much closer to a low-pressure model since there is little sign of a 'positive check' in action. This can be seen in three respects. First of all, the long-term relationship between real wages and life expectation is quite different

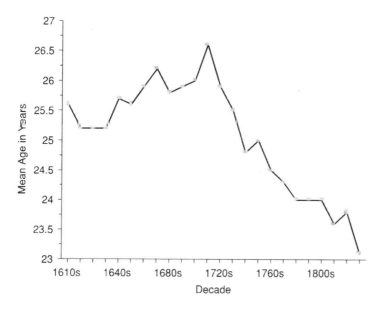

Figure 2.4 Female ages at marriage in English family reconstitutions
Source: Wrigley *et al.* 1997: Table 5.3

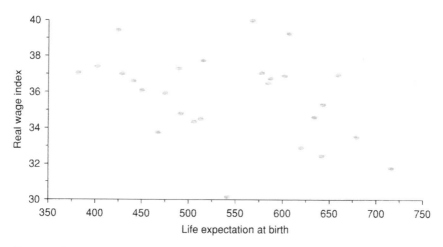

Figure 2.5 Real wages and life expectation at birth
Source: see text

from what we should expect in a high-pressure regime. For a while in the seventeenth century mortality rose as living standards fell, but this was unusual, and Figure 2.5 shows that mortality was generally more severe in decades when real wages were high. Second, since the 'high-pressure' model implies that a large proportion of the population lived on the margin of subsistence even in years with 'normal' food prices, any appreciable short-term rise in prices should result in a detectable increase in death rates. But Wrigley and Schofield's ABP results show that years with exceptionally high prices were as likely to have below- as above-average mortality.[8]

The final piece of evidence comes for a comparison of general mortality levels with those of the elite. If, as in a high-pressure regime, poverty forces up mortality among a large section of the population then elite mortality should be substantially below the general level. But a comparison of the Wrigley–Schofield series with estimates for the British peerage shows that for most of the period this was not the case; indeed peerage mortality was sometimes the higher of the two (Livi-Bacci 1991: Chapter 4). These findings raise the question of what produced such variations if it was not the standard of living. The question has yet to be fully resolved, but it seems clear that exposure to infection was the major variable. This emerges starkly from a consideration of geographical differences, which were substantial and dominated by the contrast between mortality levels in large towns and cities – London above all for most of the period – and those in smaller settlements elsewhere. Since members of the elite spent more of their time in the capital, and other large centres, than did the general population, it is likely that this made a major contribution to their elevated mortality.

Adaptation and maladaptation

We began by defining four instances of adaptation based on permutations of two adaptive criteria, Malthusian and Darwinian, and two adaptive levels, macro and micro. The demographic and economic data demonstrate that, relative to its European neighbours, England's demographic regime was adaptive at the macro-level on both criteria. England's near trebling of population between the mid-sixteenth and late eighteenth centuries compares with a west European average that was probably around 65 per cent (Wrigley 1985). Holland may have rivalled England's performance in the first growth cycle and Ireland in the second, but no other country in the region could match it over the period as a whole. The proximate explanation is not far to seek. There is no evidence that overall fertility was unusually high – indeed fertility within marriage was rather low by west European standards (Wilson 1984) – and immigration was not a major contributor, so long-term death rates must have been lower than elsewhere.[9] Economically, there was a long-term expansion in per capita output which was large in contemporary terms. At its root was an increase in the productivity of agricultural labour which, well before industrialisation, meant that the total population for every hundred directly dependent on agriculture rose, on Wrigley's estimates, from 132 in 1550 to 219 in 1750 (Wrigley 1985). As productivity rose so did real wages, and by the late eighteenth century they were the highest in Europe outside Holland.

The long-term performance is striking, but should not obscure the existence of three sub-periods with very diverging characteristics. Over the first century population growth exceeded the ability of the economy to absorb additional labour, and living standards fell until a reduced nuptiality and higher mortality brought growth to an end. A century of demographic stagnation followed during which economic consolidation enabled living standards to recover, and the basis was laid for renewed population growth, after 1750, which for some time proved compatible with stable or rising living standards. England's process of demographic adaptation relied on a national variant of the so-called North West European Marriage Pattern (NWEMP) of delayed, and less than universal, marriage. Historically, the origins of NWEMP remain obscure, but functionally it rested on a pattern of nuclear family households and neo-local marriage. Couples thus had to establish a new, structurally independent, household on marriage, rather than joining an existing one. Accumulating the economic wherewithal for this could take some years, if it was achieved at all (Smith 1981; Hajnal 1983).

In England, the fruits of labour force participation seem to have contributed more to this accumulation than did inheritance, and this fostered a homeostatic relationship between labour markets, birth rates and population growth (Schofield 1989). As population pressure eroded living standards, acquiring 'the means to marry' became more difficult, and so birth rates fell. Elsewhere in western Europe inheritance may have been more important, and fertility correspondingly less sensitive to labour market conditions, but it is likely that the region's birth rates generally were below those of major population centres elsewhere, and that living standards and mortality both benefited as a result (Wrigley 1987b). But if the region's mortality advantage reflected its higher living standards, it does not necessarily follow that England's advantage relative to its neighbours is to be explained in the same terms. Malthusian theory, and the theory of the low-pressure demographic regime, imply that there is a threshold above which improvements in living standards make little difference to mortality, and it is quite possible that western Europe as a whole was above this threshold for most of the period.[10] If this is so then large-scale intra-regional mortality differences are likely to have resulted from differences in exposure to infection, and England's advantage may have lain in its geographical isolation from the mainland, and in the very limited extent of domestic warfare with all the epidemics and disruption which this could bring in its train (cf. Mielke, Chapter 4 in this volume).

England's economic performance reflected a number of factors. First, demographic expansion was accompanied by the long-term maintenance, or even expansion, of capital stocks, particularly stocks of human capital and domestic animals. Population growth, though substantial, was thus not so rapid that potential investment resources were simply 'eaten up' by the need for additional subsistence, and nuptiality is likely to have made the crucial contribution in this respect. Technological innovation also played an important part and seems to have maintained agricultural yields per acre (Campbell and Overton 1991; Overton 1991). But the major factor in productivity growth during the 'consolidation' period was probably the development of geographical specialisation in a newly integrated national economy, a process which allowed each region to specialise in the activities for which its natural endowments best suited it (Kussmaul 1990).

Occupation specialisation also contributed to output growth, particularly during the second wave of population growth. Renewed urbanisation played an important part by providing employment for the rural population surplus. Potentially inefficient workers were removed from the agricultural labour force, and the increased number of consumers further stimulated productivity gains. Non-agricultural occupations also proliferated in the countryside and enhanced productivity by the provision of more efficient specialist services than those cultivators could provide for themselves (Wrigley 1986b). The rural economy seems to have been less successful at generating additional employment during the first wave of population growth (Goldstone 1986), and urbanisation was also on a more modest scale outside London so that the capital absorbed a grossly disproportionate share of the 'marginal' population. In doing so it doubtless ameliorated their disruptive effects on the countryside and smaller town, but London's expansion also made a more positive contribution to long-term output growth. The capital played an essential permissive role as an entrepôt and communication centre in the integration of the English national market, but it also directly stimulated the process. The size of the London market went a long way towards remedying the demand failures endemic to an organic economy and thus acted as the starter motor, if not the actual engine, of economic growth (Wrigley 1967; Fisher 1971; Chartres 1986).

On the micro-level, however, the adaptive foundations of these macro-outcomes were less secure. On the Malthusian criterion, gross inequalities of wealth and income notwithstanding, it is overwhelming likely that England's higher average incomes meant that actual living standards were also higher for most people. But it seems equally clear that, in Darwinian terms, macro-adaptation was based on fertility and migration propensities that were often detrimental to an individual's evolutionary fitness. The most obvious example is the seventeenth-century nuptiality decline. This underlay England's contemporary demographic adaptation but, given the very low rates of contemporary extramarital fertility (Adair 1996), effectively required many women to commit genetic suicide. And this is only the most striking instance of a more general problem. Late marriage and extended birth intervals meant that the completed fertility of women who did marry remained low by the standards of 'natural fertility' populations. This had a range of benign macro-consequences, as well as enabling the individuals in question to enjoy higher living standards than they would otherwise have done. But it is far from clear that they thereby had more surviving offspring than would have been the case had they married earlier and thereby 'traded off' enhanced economic welfare for a longer reproductive career. The absence of any convincing evidence for a strong intra-national relationship between mortality and living standards rather suggests the opposite.

The level of fertility within marriage raises a similar question. Maternal investment in prolonged breast-feeding reduced both fertility and infant mortality, but it did not necessarily boost survival sufficiently to compensate for the smaller number of children born. In fact data from elsewhere in Europe suggests that the opposite may have been true. Pre-twentieth-century Germany displayed large

Table 2.3 Marital fertility and child survival in German village studies

Region (1)	N (2)	Infant mortality+ (3)	Survival to age 10* (4)	Index of marital fertility # (5)	(4) × (5) (6)
Baden	4	0.234	0.65	0.80	0.52
Württemberg	1	0.300	0.62	0.86	0.53
Bavaria	3	0.321	0.64	0.93	0.60
Waldeck	4	0.188	0.67	0.79	0.53
East Frisia	2	0.145	0.77	0.67	0.52

Notes: + All years of birth
       * Years of birth –1849
       # 1750–1849

Source: Knodel 1988: Tables 3.1, 3.3, 10.1

regional variations. Bavaria, in particular, had low levels of breast-feeding, while East Frisia, in the north-west, seems to have resembled England, and other regions were intermediate between the two. The figures in Table 2.3 report results from village reconstitution studies from periods between the later eighteenth and late nineteenth centuries. They show, as we might expect, that both marital fertility and infant mortality were higher in Bavaria, but the region's excess fertility persists even when the index is adjusted for survival to age 10, and the north-west enjoys no advantage over the other regions.[11] This is because high infant mortality tended to be associated with reduced mortality later in childhood, presumably because artificially fed infants who survived their first year enjoyed a much higher level of subsequent immunity. A strategy of concentrating maternal resources on a smaller number of infants thus failed to boost Darwinian fitness in practice because infectious disease was no respecter of the level of investment in a given individual.

Mortality from infectious disease is also implicated in the migration paradox. This centres on the role of London, a role which, as we have seen, was economically crucial. This was partly a matter of the country's physical geography which enabled produce from a very wide area to reach it by water and thus got round the problems of land haulage. But geography was not everything. For most of the seventeenth and eighteenth centuries London recorded a steady surplus of burials over baptisms, and it could not have survived without a continuous flow of immigration. During the first wave of population growth it effectively 'drained off' much of the rural population surplus, but this effect, benign at the macro-level, was based on a propensity to migrate which was seriously maladaptive in micro-Darwinian terms. Since mortality was so much higher in London than elsewhere, there can be little doubt that most rural immigrants would have contributed more to the gene pool of subsequent generations had they remained where they were, and the peerage data indicate that riches were no protection in this respect.[12] If we take the long view of evolution, the human propensity to invest in offspring 'quality', and to seek out new and unfamiliar environments, has doubtless made a powerful contribution to the Darwinian success of our species. But our phylogenetic inheritance was moulded in an environment where infectious disease mortality was not a major demographic force, and in this period it elicited behaviour which was maladaptive in terms of individual fitness.

Notes

1 The analysis in this chapter draws extensively on Wrigley's arguments. For a full elaboration of the 'organic economy' concept, and its implications for English economic history, see Wrigley (1988, 1991a, 1991b), which focus particularly on the crucial issue of energy inputs.

2 This argument has been developed particularly by the economic historian Jan de Vries. For a summary view see de Vries (1976: Chapter 6). For an extended discussion of the problem as it affected economic growth in the Dutch republic see de Vries and van der Woude (1997).

3 See Grigg (1980) for an extended review of the limits to growth in pre-industrial agriculture. Overton (1996) shows how some of these limits were overcome in the English case.

4 Where inheritance was 'partible', and property was divided among a number of heirs, the consequences of population growth were more likely to be the proliferation of smallholdings and the spread of a more egalitarian misery.

5 On the eve of the railway age it was estimated that wheat would double in price if hauled more than 65 miles by land, and other grains after 50 miles. Cheaper, bulkier commodities fared even worse, so that potatoes doubled in price only 10 miles from their point of origin (Grigg 1992: 71–3).

6 The gross reproduction ratio is equivalent to the average number of daughters born to a woman who survives to the age of 50. The expectation of life at birth is simply the average length of life of those born alive.

7 The ABP results are reported in Wrigley and Schofield (1981) and those from the GIP analysis in Wrigley et al. (1997). For the most part the latter refine rather than replace the original conclusions. The main new findings are that fertility within marriage rose somewhat in the later eighteenth century and that there was a substantial shift in the age pattern of mortality beginning in the years around 1700. See Ruggles (1999) for a detailed evaluation of the reconstitution data to which GIP was applied, and, for a more critical standpoint Razzell (1994).

8 See Schofield (1985) for a review of the relationship between prices and mortality in England. A weak association detectable between high prices and above average mortality in the following year was probably due to increases in exposure to infection as a result of increased vagrancy and crowding. For a review of evidence for such effects, and the general problem of short-term price effects on mortality, see Landers (1993: 14–22).

9 The existing direct evidence mostly supports this deduction. National mortality series, comparable to those for England, become available around the middle of the eighteenth century for Sweden and France. At that time Swedish level was comparable to England's and France's much higher (Vallin 1991). Results from a large number of local studies, assembled by Flinn, suggest that infant and child mortality outside London was generally mild by west European standards (Flinn 1981).

10 See Kunitz and Engerman (1993) for a review of historical relationships between mortality and income change; for the English case see Kunitz (1987).

11 The index of marital fertility Ig was constructed by the demographer Ansley Coale and expresses observed levels as a ratio to those of the North American Hutterites whose marital fertility is the highest reliably recorded. For its construction and application see Coale and Watkins (1986).

12 Metropolitan excess mortality was primarily concentrated in infancy and childhood but also affected young adult migrants from the countryside who lacked resistance to metropolitan diseases such as smallpox (Landers 1993).

References

Adair, R. (1996) *Courtship, Illegitimacy and Marriage in Early Modern England*, Manchester: Manchester University Press.

Blackmore, S. (1999) *The Meme Machine*, Oxford: Oxford University Press.

Bowler, P. J. (1989) *Evolution: the history of an idea*, Berkeley: University of California Press.

Campbell, B. M. S. and Overton, M. (eds) (1991) *Land, Labour and Livestock: historical studies of European agricultural productivity*, Manchester: Manchester University Press.

Chartres, J. (1986) 'Food consumption and internal trade', in Beier, A. L. and Finlay, R. (eds) *London 1500–1700*, London: Longman, pp. 168–96.

Clay, C. G. A. (1984) *Economic Expansion and Social Change: England 1500–1700*, Cambridge: Cambridge University Press.

Coale, A. J. and Watkins, S. C. (1986) *The Decline of Fertility in Europe*, Princeton, NJ: Princeton University Press.

Dawkins, R. (1976) *The Selfish Gene*, Oxford: Oxford University Press.

de Vries, J. (1976) *The Economy of Europe in an Age of Crisis, 1600–1750*, Cambridge: Cambridge University Press.

de Vries, J. and van der Woude, A. (1997) *The First Modern Economy: success, failure, and perseverance of the Dutch economy, 1500–1815*, Cambridge: Cambridge University Press.

Fisher, J. (1971) 'London as an engine of economic growth', in Bromley, J. S. and Kossmann, E. H. (eds) *Britain and the Netherlands, vol. 4: metropolis, dominion and province*, The Hague: Martinus Nijhoff, pp. 3–16.

Flinn, M. W. (1981) *The European Demographic System 1500–1820*, Brighton: Harvester.

Goldstone, J. (1986) 'The demographic revolution: a re-examination', *Population Studies*, 40, 5–33.

Goldstone, J. (1991) *Revolution and Rebellion in the Early Modern World*, Berkeley and London: University of California Press.

Grigg, D. (1980) *Population Growth and Agrarian Change*, Cambridge: Cambridge University Press.

Grigg, D. (1992) *The Transformation of Agriculture in the West*, Oxford: Blackwell.

Hajnal, J. (1983) 'Two kinds of pre-industrial household formation system', in Wall, R. (ed.) *Family Forms in Historic Europe*, Cambridge: Cambridge University Press, pp. 65–104.

Knodel, J. (1988) *Demographic Behaviour in the Past*, Cambridge: Cambridge University Press.

Kunitz, S. J. (1987) 'Making a long story short: a note on men's heights and mortality in England from the first through the nineteenth centuries', *Medical History*, 31, 269–80.

Kunitz, S. J. and Engerman, S. L. (1993) 'The ranks of death: secular trends in income and mortality', in Landers, J. (ed.) *Historical Epidemiology and the Health Transition*, Canberra: Australian National University. Supplement to volume 2 of *Health Transition Review*, pp. 29–46.

Kussmaul, A. (1990) *A General View of the Rural Economy of England, 1538–1840*, Cambridge: Cambridge University Press.

Landers, J. (1993) *Death and the Metropolis: studies in the demographic history of London 1670–1830*, Cambridge: Cambridge University Press.

Landers, J. (1994) 'Stopping, spacing and starting: the regulation of fertility in historical populations', in Dunbar, R. I. M. (ed.) *Human Reproductive Decisions: biological and social perspectives*, London: Macmillan, pp. 180–206.

Laslett, P. (ed.) (1973) *The Earliest Classics*, Farnborough: Gregg.

Livi-Bacci, M. (1991) *Population and Nutrition: an essay on European demographic history*, Cambridge: Cambridge University Press.

Overton, M. (1991) 'The determinants of crop yields in early modern England', in Campbell, B. M. S. and Overton, M. (eds) *Land, Labour and Livestock: historical studies in European agricultural productivity*, Manchester: Manchester University Press, pp. 285–321.

Overton, M. (1996) *Agricultural Revolution in England: the transformation of the agrarian economy 1500–1850*, Cambridge: Cambridge University Press.

Razzell, P. (1994) *Essays in English Population History*, London: Caliban.

Ruggles, S. (1999) 'The limitations of English family reconstitution: English population history from family reconstitution 1580–1837', *Continuity and Change*, 14, 105–30.

Schofield, R. S. (1985) 'The impact of scarcity and plenty on population change in England, 1541–1871', in Rotberg, T. I. and Rabb, T. K. (eds) *Hunger and History: the impact of changing food production and consumption patterns on society*, Cambridge: Cambridge University Press, pp. 67–94.

Schofield, R. S. (1989) 'Family structure, demographic behaviour and economic growth', in Walter, J. and Schofield, R. S. (eds) *Famine, Disease and the Social Order in Early Modern Society*, Cambridge: Cambridge University Press, pp. 279–304.

Smith, R. M. (1981) 'Fertility, economy and household formation in England over three centuries', *Population and Development Review*, 7, 595–622.

Vallin, J. (1991) 'Mortality in Europe from 1720 to 1914: long-term trends and changes in patterns by age and sex', in Schofield, R. S., Reher, D. and Bideau, A. (eds) *The Decline of Mortality in Europe*, Oxford: Clarendon, pp. 38–67.

Weir, D. (1984) 'Rather never than late: celibacy and age at marriage in English cohort fertility, 1541–1871', *Journal of Family History*, 44, 27–47.

Wilson, C. (1984) 'Natural fertility in Pre-Industrial England, 1600–1799', *Population Studies*, 38, 225–41.

Wrigley, E. A. (1967) 'A simple model of London's importance in changing English society and economy 1650–1750', *Past and Present*, 37, 44–70.

Wrigley, E. A. (1983) 'Malthus's model of a pre-industrial economy', in Dupâquier, J., Fauve-Chamoux, A. and Grebenik, E. (eds) *Malthus Past and Present*, London: Academic Press, pp. 111–24.

Wrigley, E. A. (1985) 'Urban growth and agricultural change: England and the continent in the early modern period', *Journal of Interdisciplinary History*, 15, 683–728.

Wrigley, E. A. (1986a) 'Elegance and experience: Malthus at the bar of history', in Coleman, D. and Schofield, R. S. (eds) *The State of Population Theory*, Oxford: Blackwell, pp. 46–64.

Wrigley, E. A. (1986b) 'Men on the land and men in the countryside: employment in agriculture in early-nineteenth-century England', in Bonfield, L., Smith, R. and Wrightson, K. (eds) *The World We Have Gained: histories of population and social structure*, Oxford: Blackwell, pp. 295–336.

Wrigley, E. A. (1987a) 'The classical economists and the industrial revolution', in Wrigley, E. A., *People, Cities and Wealth*, Oxford: Blackwell, pp. 21–45.

Wrigley, E. A. (1987b) 'No death without birth: the implications of English mortality in the early modern period', in Porter, R. and Wear, A. (eds) *Problems and Methods in the History of Medicine*, Beckenham, Kent: Croom Helm, pp. 133–50.

Wrigley, E. A. (1987c) 'Urban growth and agricultural change: England and the continent in the early modern period', in *People, Cities and Wealth*, Oxford: Blackwell, pp. 157–93.

Wrigley, E. A. (1988) *Continuity, Chance and Change: the character of the industrial revolution in England*, Cambridge: Cambridge University Press.

Wrigley, E. A. (1991a) 'Energy availability and agricultural productivity', in Campbell, B. M. S. and Overton, M. (eds) *Land, Labour and Livestock: historical studies in European agricultural productivity*, Manchester: Manchester University Press, pp. 323–39.

Wrigley, E. A. (1991b) 'Why poverty was inevitable in traditional societies', in Hall, I. A. and Jarvie, I. C. (eds) *Transitions to Modernity: essays on power, wealth and belief*, Cambridge: Cambridge University Press, pp. 90–110.

Wrigley, E. A. and Schofield, R. S. (1981) *The Population History of England 1541–1871: a reconstruction*, London: Edward Arnold.

Wrigley, E. A., Davies, R. S., Oeppen, J. E. and Schofield, R. S. (1997) *English Population History from Family Reconstitution, 1580–1837*, Cambridge: Cambridge University Press.

Wynne-Edwards, V. C. (1962) *Animal Dispersion in Relation to Social Behaviour*, Edinburgh: Oliver & Boyd.

3 British polygyny

Laura Betzig

The peers would seem, in fact, to be a highly moral group in this respect . . . if these figures can be believed.

T. H. Hollingsworth[1]

The British peerage

From the early seventeenth to the early twentieth century, the legitimate sons of British peers fathered 31,151 legitimate children. They made 314 bastards. In other words, as far as T. H. Hollingsworth was concerned, about 1 per cent of their young were conceived in sin. In his words: 'It was only the odd peer who had a large illegitimate family.' The first Earl Ferrers (1650–1717) was one, with some 30 bastards and 27 legitimate children, but he was an unusual man. Why? Not, according to the demographer, because the data are inaccurate or incomplete. It was, after all, 'much to the mother's advantage to have her child recognized and there was no strong reason for the father to ignore his bastard except in puritan times.'[2]

Maybe not. But nobles have often been in the habit of ignoring their bastards. Sir Ronald Syme, in a classic paper on 'Bastards in the Roman Aristocracy', could not find any. Not, he said, because they were not there, but because they were taken for granted. Some Roman emperors, senators and other rich men – from Lucretia's rape at least – took advantage of poorer men's daughters and wives. The Latin sources are full of that sort of thing. Rome's first emperor Augustus, for instance (says Suetonius) 'harboured a passion for deflowering girls – who were collected for him from every quarter, even by his wife.' Augustus' successor Tiberius (says Tacitus) had his slaves collect women: they scoured Rome for pretty, young aristocrats then 'rewarded compliance, overbore reluctance with menaces and – if resisted by parents or relations – kidnapped their victims and violated them on their own account.' Worse, Tiberius' successor Caligula was (to Cassius Dio) 'the most libidinous of men', having among other things 'seized one woman at the very moment of her marriage, and had dragged others from their husbands.'[3] Even more critical, for the production of bastards, was slavery. There were millions of slaves in imperial Rome; roughly one-third of the slaves commemorated on Roman tombs were women; and sexual access to slave women was taken for granted by masters but taken at risk by other men. Slave women had few jobs to do, but were rewarded for bearing children; early in the empire, roughly half a million new slaves were added each year, and most of those were probably the children of slave women. *Filii naturales* fill Latin literature, Latin law and the Latin inscriptions. Maybe most incriminating of all, homeborn slaves, or *vernae*, were given the same wet nurses, pedagogues, *peculium* (or allowance) and legacies as masters' legitimate daughters and sons; they were often freed, and often freed young – before having a chance to perform any useful function. Besides, they were saved places – along with their children, and even their children's children – in their masters' family tombs.[4] Nevertheless Latin bastards, like other bastards, were unambiguously bastards. The *lex Papia Poppaea* of AD 9, for instance, explicitly kept illegitimates off the official birth register.[5]

The same sort of thing went on in the Middle Ages, in spite of the church. According to Lambert of Ardres' *Historia Comitum Ghisnensium*, for instance, the big bedroom at the core of the castle where the early-thirteenth-century counts of Guines laid down had direct access to their servant girls' quarters, to the rooms of 'adolescent girls' upstairs, and to what Lambert called the 'warming room' (a 'veritable incubator for the suckling infants'). Lambert's benefactor, Count Baldwin, was buried with 23 bastards in attendance – alongside 10 living legitimate daughters and sons. 'From the beginning of adolescence until his old age, his loins were stirred by the intemperance of an impatient libido,' as Lambert remembered; 'very young girls, especially virgins, aroused his desire.'[6] Throughout the Middle Ages, extramarital sex was taken for granted: in female-biased sex ratios at well-to-do houses and even monasteries; in religious and secular laws; in papal dispensations; in registers from inquisitions; in lais, pastourelle and fabliaux; and in scores of chronicles like Lambert of Ardres'. These are tips of the icebergs: counts, let alone clerks, were not in the bastard counting game. The point – in medieval Europe, in Rome, and long before – was *not* to count them. The point was to keep claims on one's estate *down*.[7]

In his late-nineteenth-century classic, *The History of Human Marriage*, Edward Westermarck made it clear that 'the general rule is undoubtedly that one of the wives holds a higher social position than the rest or is regarded as the principal wife; [with] the children or sons or the eldest son of the first wife taking precedence over those of the later wives in inheritance or succession or otherwise.'[8] That is so of most of the world's empires. In Mesopotamia, where civilization began, noble men with one wife had sexual access to hundreds of slaves; the 'Urukagina Reform Document', one of the oldest tablets on earth, complains that 'the houses of the palace harem and the fields of the palace harem, the houses of the palace nursery and the fields of the palace nursery crowded each other side by side.' In Egypt, pharaohs kept one 'Great Wife', and filled women's rooms with hundreds of subjects requisitioned as tribute. In India – where the size of the royal seraglio was thought to be 16,000 in the sixth century BC – *rajas* kept just one 'chief queen' at the top. In China, handbooks recommended early – by the start of the Zhou dynasty (771–256 BC) – that kings keep a queen, along with 3 primary consorts, 9 wives of second rank, 27 wives of third rank, 81 'concubines' and other women who came to number in the thousands. As always, 'palace agents used to scour the empire' for pretty women, and took them wherever they found them.[9]

In England, as in imperial Rome, most of medieval Europe, and almost everywhere else, a rich man confined himself to one *wife* at a time. Her children alone were his legitimate *heirs*. But having sex with a wife did not preclude having sex with other women. And fathering heirs did not preclude fathering bastards.

British peers seem to have been prolific, at least, in getting children on their wives. Their legitimate sons' mean fertility – where their wives, too, were of noble origin – was above six in the sixteenth, seventeenth and eighteenth centuries. As many as 87 peers' sons had at least 15 legitimate children; and all of 46 peers had 50 legitimate grandchildren or more. As Hollingsworth put it: 'There is a certain appeal in founding a great dynasty and seeing descendants spread all over the country and even all over the world.'[10] Not all of that seed need be legitimate seed.

The bastardy ratio

The Cambridge Group for the History of Population and Social Structure has amassed data from over four hundred sixteenth- to nineteenth-century parishes, on a number of variables including bastardy. After a modest start at around 2 per cent, the illegitimacy ratio climbs, with several jags, to a high of almost 6 per cent in 1830 when the ecclesiastical records run out, then to a new high of 8 per cent in the twentieth-century civil registration era. Overall, the peaks and valleys in bastardy correspond 'uncannily' to those in legitimate fertility. In Peter Laslett's words, 'illegitimate fertility is a sub-set, so to speak, of fertility as a whole, always a small part of it and varying as overall fertility varied.'[11]

There are a number of problems involved in interpreting these results. For one thing, some parish clerks were better at counting bastards than others. There is a 'pronounced tendency' for counties to maintain a high or low ratio from decade to decade; but there is no way to know how much such variation reflects real differences in illegitimacy, and how much it reflects differences in reporting. In many parishes, decades go by without a single bastard being recorded, though it is unlikely that not one bastard was born. For another thing, some bastards are more likely to be counted than others. For instance, children who died in infancy were more likely to be left off the records, and bastards were more likely than other children to die young. Unwed mothers often ran away to London, where non-registration of births ran as high as 70 per cent. And many marriages were backdated to cover a pre-nuptial pregnancy up. Last, bastards born to married women – always tough to count, but estimated to be as high as 20–30 per cent in a recent study of the 'Liverpool flats' – are not included in these results.[12]

In an earlier study of bridal pregnancy, P. E. H. Hair looked at a sample of 3786 marriages drawn from registers of 77 parishes in 24 English counties for the period 1550 to 1820. Of the 1855 brides he was able to 'trace to maternity'

– that is, for whom the birth of a child was also uncovered in a register – roughly one-third gave birth within eight and a half months of marriage. Hair took this to be a serious underestimate of bridal pregnancy because, among other things, baptism (the more common date of entry) tended to take place several days or even weeks after birth. He concluded that 'the figure for pregnancy among traced brides begins to approach half of the sample.'[13] A decade and many data later, Laslett's conclusion was similar. Conservatively, roughly one-fifth of all first conceptions in England may have been extramarital; but the proportion 'was usually more like two-fifths'; and it 'could easily reach three-fifths.'[14] Other evidence suggests rates like these may extend back several centuries. As Richard Smith has pointed out, the large numbers of childwite and leyrwite payments recorded in manor court rolls 'show just how widespread and lacking in social stigma pregnancy outside of wedlock could be.'[15]

Who mothered these bastards? Pregnant brides were often poor. In, for instance, sixteenth- and seventeenth-century Terling, 61 bastards were born to 50 mothers. Of these mothers 9 were of 'middling' status – of yeoman, husbandman or craftsman origin; and the status of another 22 mothers was 'low' – mostly servants' or labourers' daughters. Though the status of the other 19 mothers was unclear, 'their very obscurity in the records would suggest that they were of low social status. . . . Many may have been servants.' In seventeenth-century Lancashire 24 per cent, and in seventeenth-century Essex 61 per cent, of the mothers of bastards were servants; and no mother ranked higher than a yeoman's daughter. In sixteenth- to nineteenth-century Colyton, Alcester, Aldenham and Hawkshead, servants and labourers are overrepresented among bastards, husbands of bastards and fathers of mothers of bastards. In eighteenth- and nineteenth-century Banbury, labourers and servants are the only overrepresented group among 'fathers of mothers of bastards'; while in eighteenth- and nineteenth-century Hartland, no one occupational group was particularly bastardy prone. In earlier evidence from manor court rolls in thirteenth- and fourteenth-century Suffolk, childwite payments are disproportionately represented among the lower social strata. And as late as 1883, an investigation by the Registrar General in Scotland showed that almost half of illegitimate births were to domestic servants. It fits that when in 1851 figures are available at last for proportions of servants by county, they show a significant correlation with bastardy. 'A county with more servants had more bastards.'[16]

Who fathered these bastards? That is the trickier, crucial question. Even in the parish records, it is clear that bastards' putative fathers were higher-ranking overall than bastards' mothers. In, for instance, seventeenth-century Essex, at least 23 per cent of fathers named 'were in a magisterial postition vis-à-vis the mothers; that is, they were the masters of the girls, members of their master's kin, or gentlemen.' The same holds for 14 per cent of seventeenth-century Lancashire fathers. Another 28 per cent of Essex and 8 per cent of Lancashire fathers were fellow-servants of bastards' mothers. In sixteenth- and seventeenth-century Terling, 13 putative fathers were of 'middling' status, 15 were 'low' and another 21 'obscure'. That means 'a higher proportion of fathers than of mothers was drawn from the upper social categories.'[17] Even more important, 'paters' (putative fathers) were not necessarily 'genitors' (genetic fathers). There are plenty of hints of this. An eighteenth-century Paisley Central Library manuscript attributed to Andrew Crawfurd, 'Cairn of Lochwinnoch Matters', reads:

> Lord Crawfurd had a highland lass or mistress, a bonnie nymph, from Arran. Auld Auchinhane reported my Lord bairned her and [being tired of her] he garred Willie Orr, his servant to kipple with her, and take the wyte of his lordship's wane. Therefore Lord Crawfurd set Willie up as a farmer . . . and tochered Jean, the issue of his amour.[18]

To the extent that arrangements like these could be made, parish registers – even where they have plenty to say about putative paternity – say little about genetic paternity. Household size may get at that more directly.

Household size

John Graunt, doing the first British demography in 1662, guessed households typically held a husband, a wife, three children – and three servants. John Hajnal, who identified the north-west European marriage pattern in 1965, included several servants in each household, and estimated around 17 per cent of the population was probably 'in service' at once. British servants show up in medieval documents as *servus, serviens, ancilla, garciones* and the more ambiguous *famulus*; Richard Smith found them, as early as 1377, in 20 per cent of households in Rutland. And Alan MacFarlane speculated that England's 'particular marital, demographic, political and economic systems' extend back as far as, or farther than, most of the records go – back to thirteenth-century Britain, if not to Tacitus' first-century *Agricola* and before.[19]

In short, the *mean* in British household size has stayed pretty much the same – at just under five, from the earliest parish records until very recently. At the same time, the *variance* has declined.[20]

Dramatically: by the late Middle Ages, great magnates' households consistently numbered in the hundreds. According to Stowe's pre-Civil War survey of London, 'in every great house when occupied there would be 500 knights and men with their horses and their grooms and their cooks, bakers, brewers, valets, footmen, stable-boys, blacksmiths, armorers, makers and menders.' Richer servants – the sons of lesser nobles and gentry – waited tables and carved, hunted and hawked, and honed their fighting skills with their betters. The sons of many more poor men worked behind or out of doors. As late as the fifteenth-century reign of Henry VI, the Earl of Salisbury had 500 men on horseback; the Duke of York kept 400 men at Baynard's Castle; and the Dukes of Exeter and Somerset employed 800 men each.[21]

Then suddenly, in the sixteenth and seventeenth centuries, the numbers started to drop. In the mid-fifteenth century, the Earl of Warwick went to Parliament with 600 servants; George Duke of Clarence kept 299 men in livery; and the mere deputy-steward of Kendal in Westmorland had a retinue of 290. Less than a hundred years later, in the early sixteenth century, the Earl of Northumberland had just 171 in his household and the Bishop of Ely a hundred – though Cardinal Wolsey still kept a checkroll of 422. By the mid-sixteenth century, Sir Thomas Lovell, the Earl of Rutland and the Earl of Derby kept just around 100 on their household staffs. And by the end of the seventeenth century, in 1693, Nottingham's household held just 49 servants. The numbers continued to slip. A typical eighteenth-century household numbered in the lower tens. In 1734, the Duke of Newcastle kept just 40 servants at Claremont; the Duke of Norfolk had 30 at Arundel in 1775; Lord Gage had around 30 in 1774; Lord Hardwick 30 in 1778. Especially telling is the shrinkage on individual estates. In the 1690s, the Duke of Somerset kept a staff of more than a hundred; in the early eighteenth century, it was down to around 55; and though in 1587 the Earl of Derby kept 118 servants at Knowsley, in 1702, his successor got by with a staff of just 38. Bishop Burnet among others, in 1713, found it 'a happiness to the nation that the great number of idle and useless retainers that were among noblemen anciently is much reduced.' The end came with the Second World War, when servants virtually disappeared. In 1939 the Marquis of Bath still, extraordinarily, had 40 live-in domestics; 40 years later, his successor made do with 2, both Spanish.

Gregory King, in his *Scheme of the income & expence of the several families of England* of 1688, counted 160 families under the heading of 'Temporal Lords'; he estimated each of those households held around 40 heads. King counted 26 families of 'Spiritual Lords', at 20 heads per household; 800 'Baronets' at 16 heads apiece; 600 'Knights' at 13 heads; 3000 'Esquires' at 10 heads; and 12,000 'Gentlemen' at 8 heads each. At the other end of the social scale, King counted no fewer than 50,000 households of 'Common Seamen'; 364,000 of 'Labouring People and Out Servants'; 400,000 of 'Cottagers and Paupers'; 35,000 of 'Common Soldiers'; and a whopping 849,000 'Vagrant' families – including 'Gipsies, Thieves, Beggars, &c.' These bottom-of-the-ladder households held, respectively, 3, $3^1/_2$, $3^1/_4$, 2 and $3^1/_4$ heads.[22]

The end of service as an institution is, of course, obvious in the parish lists. Laslett estimated that in sixteenth- to nineteenth-century England, 28.5 per cent of all households included servants, and that 13.4 per cent of the total population was in service at any time. Bigger houses held bigger staffs. Servants lived in 84 per cent of 'gentlemen's' households; 81 per cent of clergymen's; 72 per cent of yeomen's; 47 per cent of tradesmen's and craftsmen's; 23 per cent of labourers' and 2 per cent of paupers'. Richard Wall counted the number of servants per 100 households from the mid-seventeenth to the late twentieth centuries and documented a steady decline. In the period 1650–1749, the number of servants per 100 households in England was 61; in 1750–1821, it was 51; in 1851, it had fallen to 33 in rural and 14 in urban regions; in 1947, it was just 2; and in 1970, the number of servants per 100 households had vanished to none. As Hajnal pointed out, the term 'servant' itself 'refers to an institution which, so far as is known, was uniquely European and has disappeared.'[23]

Household composition

Domestic servants consistently shared three traits. They tended to be young; they tended to be female; and they were almost always unmarried. Laslett, who coined the term 'life-cycle service', pointed out that 'servants in fact were, to a very large extent, young, unmarried persons – indeed sexually mature persons waiting to be married.'[24]

In 63 parish listings dated 1574 to 1820, Ann Kussmaul found just 13.4 per cent of the total population 'in service', but all of 60 per cent in service between the ages of 15 and 24. The modal age of entry into service was 13 to 16. Most outdoor servants were boys; most indoor servants were girls. In 1851, the ratio of male to female farm servants was a high 213:100, while the ratio of male to female domestic servants was as low as 13:100. Demand for domestics was especially high in towns. Populations in European cities were often sex-skewed – in part, in the words of Richard Wall, 'the product of the flooding in of domestic servants to serve in the households of wealthier masters'. London was no exception: Patrick Colquhoun, making one early-nineteenth-century estimate, put the number of menservants at 110,000, and maidservants at 800,000.[25]

Servants were usually unmarried, especially girls. In a Tudor letter to Honor Lisle (the viscount's wife), a Lady Mary Jeromymne becomes ineligible for service when she gets engaged. The writer says, 'she will marry and serve no more'. Similarly, two centuries later, Elizabeth Purefoy, matron of a Buckinghamshire manor, wrote in May 1739 to her goddaughter, "tis not my dairy maid that is with child but my cookmaid, and it is reported that our parson's maid is also with kinchen by the same person who has gone off & showed them a pair of heels for it.' She adds, with a common lack of concern, 'if you could help me to a cookmaid I may be delivered from this' – servants with child by outsiders were typically turned out in disgrace.[26] In the eighteenth century, the Duke of Bedford's upper serving men had wives and homes of their own; but his lower serving men, and all of his serving women, were unmarried and slept in his house – the women upstairs in the attics, and the men in the cellars downstairs. Jonas Hanway, in the mid-eighteenth century, noted that 'few people have humanity and patriotism sufficient to entertain married servants.' As David Hume is often quoted for having said: 'all masters discourage the marrying of their male servants and admit not by any means the marrying of the female who are then supposed altogether to be incapacitated for their service.'[27]

Some evidence suggests that Englishmen might have preferred servant girls who were not just unmarried, but chaste. Housemaids, often quartered in the attic, usually slept two to four to a room. Occasionally, their windows were barred. Christopher Smout quotes a nineteenth-century Scot, Alfred List, who wrote: 'the cry everywhere is "No followers!" as if the young serving girl had no title to love or to be loved.' List went on with this advice: 'before you retire to rest lock the doors with your own hand, and deposit the keys upon your toilet table, otherwise, while enjoying the blessings of repose above you may not know what junketings are going on below.'[28]

The classicist Susan Treggiari, in a paper called 'Jobs for women', pointed out that while the inscriptions on Latin tombs tagged male slaves with a wide variety of jobs, female slaves seem to have had little to do in imperial Rome. There are no job titles for women who worked outdoors; and women rarely seem to have worked in the public parts of houses. Where jobs are attested at all, they seem to have waited on other women. They were dressers, hair dressers, clothes menders, clothes folders, massagers, midwives, wet-nurses, spinners and weavers. But 'there are many women in the *columbaria* whose jobs are not known.'[29]

There are parallels in England. The ranking servants on large household staffs were always male. Men were house stewards, land stewards, bailiffs, butlers, grooms, valets, clerks of the kitchen and stables, cooks and confectioners, footmen and so on. Women typically waited on women. The lady's maid dressed hair and made hats; other women cooked, tended children and kept house. Under the housekeeper – who supervised – were chambermaids, housemaids, laundry maids, dairy maids, scullery maids. Washing, picking up and dusting seem to have been their business: they stayed behind the scenes. François de la Rochefoucauld, touring England in the late eighteenth century, remarked that more than half of all servants were never seen – among others, 'maidservants in large numbers'. As far as he could see, women did the cooking, women did the cleaning, but men 'performed in the presence of guests'. And all of them were irresponsibly idle. 'English servants', he wrote, 'have practically nothing to do. They are the laziest people it is possible to meet.'[30]

How, then, did the 60 per cent of all 15–24-year-old girls in early-modern England – the hundreds of women who waited on Wolsey in the sixteenth and the scores who catered to Chandos in the eighteenth century – the 800,000 farm girls who had flocked to London to serve the rich in 1806 – pass their time? Odds are they did more than dust.

Fornication

There are, and there always have been, two ways to get bastards. One involves sex with a married woman. The other involves sex with a maid. The first is commonly known as adultery. The second is fornication.

As in any pre-DNA fingerprinting age, the data on fornication are soft. But not altogether unpersuasive. From the parish records, to legal records, to letters, diaries and magazines, to the bulk of eighteenth-century English prose, higher status men are incriminated as the fathers of bastards by lower status girls.

Bridal pregnancy may show up in a parish register after one of two events. The first is that a couple, having conceived a child, got married and raised it. This is, again, the preferred conclusion of many modern demographers. The second is that a couple, having conceived a child, got another man to raise it. This was an ancient option. William of Malmesbury, for instance, notes with disgust in his twelfth-century *Gesta Regni* that on the eve of the Norman conquest the English nobility were 'given up to luxury and wantonness'. He adds: 'There was one custom, repugnant to nature, which they adopted; namely, to sell their female servants, when pregnant by them and after they had satisfied their lust' – either to prostitution at home or to slavery abroad.[31] More often, after Lord Crawfurd's fashion, they probably married them off.

Law

When, in 1579, poor relief was mandated by Parliament – making each parish raise money to support its indigents – bastardy became one of a church court's main concerns, and establishing paternity one of its primary aims. In seventeenth-century Somerset, as in other counties, rich men were in a better position to get their lovers a pension. One Robert Sommers, having seduced his maid, had her locked up on false charges, and had a local husbandman jailed for 'truthfully' denying having fathered her child. In another case Ann Bishop, having conceived by a rich lover, named somebody else the father because 'Arthur Pulman, the now Constable of the Hundred of Martock did threaten . . . that if she did not deny that it was Robert Cribb's child he would have her whipped to the bone [and] would have her laid in the goal until she would rot.' In the Somerset county records, of men who did *not* marry their pregnant girlfriends, one in ten set them up as other men's wives. Some cuckold grooms gladly took the usual £2 to £10 paid, though others managed to get two or three times as much from a rich or desperate lover. Roger Pierce got £3 (or a heifer) plus a house rent-free for a year. Another man got promised 40s., married the girl, never got the money, and left her. Lord Gibson of Worle found his maid, Joan Salmon, to be with child. He went to the local pastor and published banns of matrimony between Joan and a Richard Carter – *without* Carter's consent.

> But as soon as . . . Carter understood thereof he came to forbid the . . . banns, but . . . Gibson with many persuasions and fair promises hath made up the match between them and hath promised . . . Carter that if he cannot receive half his brother's living for him he will give him thirty pounds himself toward the marriage of his maid.[32]

There were similar cases in other places. In Rothesay, Buteshire, Scotland, in 1661, Nancy Throw confessed she had named Alexander Bannatyne her child's father because Robert Stewart, the real culprit, 'had persuaded her that if she would confes the bairne to be his she would be drowned and he execute because he was a preacher.' Alexander had been promised 40 marks, 'a boll of vituellis', and a butt of land for taking up Robert's burden. In seventeenth-century Warwickshire, farmer Hibbert got the local Justice of the Peace (JP) to tell Richard Preston, his ex-manservant, to marry Mary Matthews, his ex-maidservant, or else 'he should be sent to a goal and put into a dungeon where he should lie and rot and that he should never come out.' Richard reluctantly married Mary, but then sued for annulment on grounds of coercion. In another, early-eighteenth-century Lancashire case, James Cash (a local constable) got Ellen Greenhalgh (his pregnant maid) to call Thomas Houghton (a bachelor blacksmith) her child's father. The truth got out when master and maid were discovered in bed together. The Quarter Sessions court dismissed the charges against Houghton, and ordered the guilty *girl* to be whipped.[33]

Sometimes, then, men got their pregnant girlfriends husbands. Other times, they just paid cash. Several notable cases came to court. Among them: Samuel Morland, the seventeenth-century inventor, sued for separation from the destitute ex mistress of a baronet who provided the cash to cover her lawsuit. Charlotte Calvert, a bastard granddaughter of Lady Castlemaine and Charles II, after separation from her husband, took up with a count who complained she had cost £2,000 cash plus £300 in household effects over two-and-a-half years. And Nancy Parsons, an eighteenth-century courtesan, was kept openly by Augustus Henry FitzRoy, the Duke of Grafton and, eventually, Prime Minister of England – at the head of whose table she proudly sat.[34]

Letters, etc.

Sometimes, bastardy came to light for the benefit of a jury. Other times, it came to light in letters, periodicals, pamphlets, memoirs, journals and diaries. The most notorious of the last is Samuel Pepys's. Pepys was a respectable man. He held posts in the navy, was Secretary of the Admiralty, and a Fellow – eventually President – of the Royal Society. To John Evelyn, his friend and fellow Royal Society member, he was 'a very worthy, Industrious & curious person'; he performed his offices with 'greate Integrity'; he was 'universaly beloved, Hospitable, Generous, Learned in many Things, skill'd in Musick' and loved the conversation of intelligent men.[35]

He loved women as well. From January 1660 to May 1669, Pepys made the mistake of writing everything down. He was obsessed with a girl named Betty Lane, seamstress at Westminster Hall. After two years of casual encounters with occasional petting over dinner, he made full contact with Betty on 16 January 1664. That night's entry notes in cryptic French: 'I by water to Westminster Hall, and there did see Mrs. Lane. . . . So home to supper and to bed, with my mind un peu troubled pou ce que fait to-day, but I hope it will be la dernier de toute ma vie.'

Pepys, being a married man, was immediately consumed with finding Betty a husband. In fact, almost a week earlier, on 9 January, he noted: 'I to Westminster Hall, and there visited Mrs. Lane. . . . So I to talk about her having Hawley, she told me flatly no, she could not love him.' But he was persistent, and tried again. On 8 February: 'I went and talked with Mrs. Lane about persuading her to Hawly, and think she will come on, I wish it were done.' The next day 'Hawly and I talked of his mistress, Mrs. Lane, and I seriously advising him and inquiring his condition, and do believe that I shall bring them together.' And again, on 29 February: 'so to Westminster Hall, and there talk with Mrs. Lane and Howlett, but the match with Hawly I perceive will not take, and so I am resolved wholly to avoid occasion of further ill with her.'

He failed to keep that resolution. Though in the end Mrs Lane had her own way – she sent Hawly away, and married Captain Martin. Pepys disapproved. Martin seemed 'a sorry, simple fellow' when Pepys first sized him up. But even a sorry, simple cuckold had to get paid off. On 16 August, Mrs Martin met Mr Pepys in a bar, told him a 'sad story' about how her husband had turned out to be poor, and ended up crying 'that she is with child and undone, if I do not get him a place.' Pepys was nonplussed! 'I had my pleasure of her, and she, like an impudent jade, depends upon my kindness to her husband, but I will have no more to do with her, let her brew as she has baked, seeing she would not take my counsel about Hawly.' He did not mean it. Two years – and two children born to Betty Martin – later, Pepys was resolved that 'the poor man I do think would take pains if I can get him a purser's place, which I will endeavour.' After another two months, in May 1666, Samuel Martin was made purser aboard the *East India London*. He was made consul at Algiers in 1674; and when he died in 1679 his widow was granted a pension by the King – possibly at the petition of his friend, Mr Pepys.[36]

About a century after Pepys left his private life in a drawer, James Boswell incriminated himself. Boswell had fallen short of his political ambitions – he was a bored Edinburgh lawyer – but lived well on the estate he got from the Laird of Auchinleck manor, his father. Boswell was richer than Pepys, and indulged his passions better. Both seem to have been deeply in love with their wives; but, where opportunity knocked, that was not enough. In all, over the course of 37 years – from 1758 to his death – Boswell's journals record affairs with 3 'women of quality', 3 of lower class, 4 actresses, a wife of the King of Prussia's guards, the mother of Jean-Jacques Rousseau's 5 illegitimate children, and more than 60 whores – in Edinburgh, London, Berlin, Dresden, Geneva, Turin, Naples, Rome, Florence, Venice, Dublin, Paris and Marseilles.[37]

Boswell was a little more philosophical than Pepys. His affairs were more matters of conscience, less matters of fact. One day on the grand tour, at the tender age of 24, he made a pilgrimage to the humble retreat of Rousseau ('Enlightened mentor! Eloquent and amiable Rousseau!'). They met at Soleure on 3 December 1764, and Boswell was keen to talk over his deepest concerns. 'Morals appear to me an uncertain thing,' he complained. 'For instance, I should like to have 30 women. Could I not satisfy that desire?' Rousseau answered in one word: 'No!' But Boswell put his case as well as anybody had:

> Consider: If I am rich, I can take a number of girls; I get them with child; propagation is thus increased. I give them dowries, and I marry them off to good peasants who are very happy to have them. Thus they become wives at the same age as would have been the case if they had remained virgins, and I, on my side, have had the benefit of enjoying a great variety of women.[38]

Some men implicated themselves; others were implicated by women. A few particularly literate courtesans felt slighted enough to try extortion. Teresia Constantia (Con) Phillips published her memoirs, an *Apology*, in the mid-eighteenth century

– hoping to extract money from lovers reluctant to have their affairs exposed. Several refused to pay up. Among them: the second son of Sir John Southcote, a landscape gardener good at getting children by other men's women ('his exploits that way recommended him to many ladies'); the rich merchant Henry Muilman, Con's second husband, who fought her for 18 years in court ('P.S. You may depend, no hush-money will be given'); and Philip Stanhope, heir to the third Earl of Chesterfield, who had asked Con Philips as a 13-year-old virgin up to his rooms to watch the fireworks, got her drunk, tied her to a chair, and initiated her.

Almost a hundred years later another courtesan, Harriette Wilson, filled her own *Memoirs* with exposés of (among others) the Dukes of Argyll, de Guiche, Leinster and Wellington, Marquesses of Anglesea, Bath, Headfort, Hertford, Sligo and, especially, Worcester. The sixth Duke of Beaufort, Worcester's father, had tried to buy Harriette off with a settlement in return for a stack of his son's love letters; she was insulted by an offer of £1,200, and the *Memoirs* followed a few years later. Harriette, when she wrote to her old loves, asked a flat £200 fee for being left out of her book; and most of them seem to have paid up. In an 8 March 1825 letter to 'Bear' Ellice, then Secretary to the Treasury, she wrote: 'people are buying themselves so fast out . . . that I have no time to attend to them.' The net, said Harriet's friend Julia Johnstone, was over £10,000.

Some kept women fared better than others. One of Harriette Wilson's contemporaries, Catherine ('Skittles') Walters, who had started out setting up pins in a Lancashire bowling alley, ended up – having been under the Duke of Devonshire's protection and 'admired' by Gladstone – being pushed about in Hyde Park, in her well-endowed dotage, by Lord Kitchener and various men of letters. As late as 1892, the eighth Duke of Marlborough left a mistress, Gertrude Campbell, £20,000 in his will. And John Clelland's eighteenth-century *Woman of Pleasure* (however close to fact his fiction) was left – having been kept by nobility, gentry and a sympathetic madam – at a tender age with a man she dearly loved and a reserve of £800. 'I saw myself in purse for a long time.' Courtesans like these kept their mouths shut.[39]

Not all the literary evidence is confessional, of course. Tracts for the edification of servants warned against getting too familiar with masters. Thomas Seaton's 1720 *Conduct of Servants in Great Families* – dedicated with unction to the Rt Hon. Daniel, Earl of Nottingham – says manservants were unlikely to be assaulted. 'The very Thought of any Temptation of this Kind coming from the Mistress of an House is exceeding shocking and unnatural.' Maidservants were another matter. 'It will sometimes happen that a Master of a great House is Young, and Wanton, and Bold, and Rakish', and not inclined to rein himself in. The good maid must keep herself hidden; she should cry out if she is found; and, if her master persists, she ought to quit. 'And if she shou'd happen to be at length a Sufferer by that impure Fire which was of her own Kindling, she is principally to be blamed.' Eliza Haywood's *Present for a Servant-Maid* of 1743 ends with the same sort of advice. The problem, as always, was one of resisting authority. 'Being so much under his Command, and obliged to attend him at any Hour, and at any Place he is pleased to call you, will lay you under Difficulties to avoid his

Importunities.' Persevere, she says, and 'perhaps, in Time', you might 'oblige him to desist'. Scold him for sinning; avoid smiling; and (again), if absolutely necessary, quit. 'How great will be your Glory, if, by your Behaviour, you convert the base Design he had upon you, into an Esteem for your Virtue!'[40]

Some masters were contractually up front about compensations they offered. One sixteenth-century Essex man spent 40*s.* on a girl to serve him by day, and another 40*s.* to serve him by night. Jonathan Swift, in his *Directions to Servants*, had this to say to the waiting-maid: 'Make him pay.' Swift starts out, 'If you are in a great family, and my lady's woman, my lord may probably like you.' Take the opportunity, he says, to be mercenary. 'Take care to get as much out of him as you can; and never allow him the smallest liberty, not the squeezing of your hand, unless he puts a guinea into it.' Swift suggests 5 guineas for a squeezed breast, and 100 guineas at least for 'the last favour' – if not a settlement of £20 a year for life. A really lucky maid could score big: Lord Townshend, who died without legitimate children, left his housekeeper and three bastards £50,000. But poorer options were always at a master's discretion. In his nineteenth-century *Function and Disorders of the Reproductive System*, it was Lord Acton's opinion that 'men who themselves employ female labour, or direct it for others, have always ample opportunity of choice, compulsion, secrecy and subsequent intimidation should exposure be probable and disagreeable.' After all, 'they can at any moment discharge her.' And did.[41]

As Sam Johnson put it in Boswell's hearing, 'a man is chosen Knight of the shire, not the less for having debauched ladies.' And, it went without saying, their maids. As Johnson went on:

> A woman who breaks her marriage vows is more criminal than a man who does it. A man, to be sure, is criminal in the sight of God: but he does not so do his wife a very material injury, if he does not insult her; if, for instance, from mere wantonness of appetite, he steals privately to her chambermaid. Sir, a wife ought not greatly to resent this. I could not receive home a daughter who had run away from her husband on that account.

Boswell and Mrs Boswell seem to have felt the same way. One night in March 1775, he came home to write: 'I was quite in love with her tonight . . . I told her I must have a concubine. She said I must go to whom I pleased. She has often said so.' He is sure his wife is 'in earnest' – as long as her husband stays affectionate and healthy. But Boswell does not commit infidelity easily. He cites biblical precedents. He considers human nature. He confesses *ad infinitum* – that 'no man was ever more attached to his wife than I was, but that I had an exuberance of amorous faculties, quite corporeal and unconnected to affection and regard'; that 'I have a warm heart and a vivacious fancy. I am therefore given to love, and also to piety or gratitude to God'; that 'I was Asiatic in imagination.'

Johnson's other good friend, Mrs Thrale, turned the same blind eye to her husband the prosperous brewer's several affairs. She ignored his bouts with gonorrhoea. She ignored his being – according to a 1773 edition of *The Westminster*

Magazine – 'more famed for his amours than celebrated for his beer', having set up Polly Hart, daughter of a dancing master, in a London bower. She ignored frequent visits at home from Jeremiah Crutchley, Mr Thrale's illegitimate son. And Mrs Thrale passed that indifference on to her daughter. As she spelled it out in her journal: 'Cecy Mostyn [the daughter] is a foolish Girl, & cannot rule her own Household – all our unfashionable Neighbours cry Shame! to see Mason her Maid with Child by the Master of the Mansion.' But Cecy, or Cecilia, didn't care. She insisted, as her mother had, that 'it is the Way'; that 'all those who understand *genteel Life* think lightly of such matters.'[42]

It was still very much the case, as Con Phillips put it to her friend Lord Chesterfield, that 'Men may be even profligate in their Amours, and none of you will dispute their being in all other respects Men of Honour.' Chesterfield himself advised his son, in a 27 March 1747 letter from London, that 'the character which most young men first aim at is, that of a man of pleasure', and that 'a *man of pleasure*, in the vulgar acceptation of that phrase, means only a beastly drunkard, an abandoned whore-master, and a profligate swearer and curser.' The rest of the letter blasts drinking, gaming and swearing – but leaves whore-mastering alone.

Still, more and more Boswells were feeling – and being made to feel – guilty. Chesterfield's friend, Joseph Addison, disapproved in the *Spectator*'s 23 October 1711 edition of a 'loose Tribe of Men' breeding all over London. They 'ramble into all the Corners of this great City', and 'raise up issue in every Quarter of the Town.' 'There is no Head so full of Strategems as that of a Libidinous Man.' Addison went on: 'When a Man once gives himself this Liberty of preying at large, and living upon the Common, he finds so much Game in a populous City, that it is surprising to consider the Numbers which he sometimes Propagates.' Addison's co-editor, Richard Steele, agreed with him. The 28 September 1711 *Spectator* had had an 'Alice Threadneedle' complain about 'Wenching' – a business by which, she said, idle men about town seduced 'raw unthinking Girls' and left them 'to Shame, Infamy, Poverty, and Disease'. Threadneedle was answered by 'A Man of Pleasure about Town' who had been sent to prison for fornication: 'Time was', he complained, 'when all the honest Whore-masters in the Neighbourhood would have rose against the Cuckolds to my Rescue.' As Dan Defoe – turned moralist at 67 – put it in *Conjugal Lewdness, or Matrimonial Whoredom*: God took one rib, not half a dozen, from Adam. ''Tis evident, one Wife to one Husband was thought best.'[43]

Literature

The sexual exploitation by masters of their maids is *the* stuff of eighteenth-century fiction. It was the stuff, eventually, of the novel. But first it was the stuff of Restoration plays. Within days of being returned to the throne, Charles II granted patents to Thomas Killigrew and William D'Avenant to set up the King's Company (eventually at the Theatre Royal in Drury Lane) and the Duke's Company (eventually at Lincoln's Inn Fields) to put on plays. Whoring is what Restoration theatre is all about. Sometimes the whores have husbands. Other times, they don't.

There is fornication in Restoration plots; it is there even in the asides. In for instance *The Parson's Wedding* (Thomas Killigrew, 1663), the Captain remarks: 'Variety is such a friend to love, that he which rises a funk coward from the lady's bed would find new fires at her maid's.' In *She Would if She Could* (George Etherege, 1668), Courtall says: 'A wife's a dish, of which if a man once surfeit, he shall have a better stomach to all others ever after.' And in *The Careless Husband* (Colley Cibber, 1705), poor Lady Easy complains: 'A vile, licentious man! Must he bring home his follies too? Wrong me with my very servant! Oh! How tedious a relief is patience!' But patience, as she presciently points out, is the 'only remedy' for a wife.[44]

Plenty of plots turn on the seduction of unmarried women. But the point is made best, perhaps, in two plays by John Dryden. In the first, *Sir Martin Mar-All* (1668), Lord Dartmouth makes love to a girl (Mrs Christian) with the help of her aunt (Lady Dupe). Mrs Christian protests that my Lord already has a wife. Lady Dupe retorts, 'there are advantages enough for you, if you will be wise and follow my advise.' Her niece gets a ruby for a kiss ('that is but one ruby for another'), a £2,000 bond for committing the 'wicked act' ('I ne'er enjoyed her niece under the rate of £500 a time'), and an innocent husband (at the cost of another £500 to his procurer) after an accomplished conception. Pepys, by the way, found this play 'the most entire piece of mirth that certainly ever was writ.'

Dryden's *Kind Keeper* was even more to the point – and suffered for it. It ran for just three nights. According to Dryden, it 'offended'; according to Robert Gould's satire *The Play House*, it was 'So bawdy it not only shamed the age,/But worse, was even too nauseous for the stage.' In the possibly edited surviving version, whores of all sorts (married and unmarried) are procured for whatever men (married and unmarried) can afford them. Like most keepers, these men expect fidelity in return for their fees. And like many kept women, Dryden's whores complain:

> 'Gainst keepers we petition,
> Who would enclose the common:
> 'Tis enough to raise sedition
> In the free-born subject woman.
> Because for his gold
> I my body have sold,
> He thinks I'm a slave for my life;
> He rants, domineers,
> He swaggers and swears,
> And would keep me as bare as his wife.[45]

Restoration plays appealed at first to men and women of the Stuart court; but in the end they were put on for the London citizens who paid to see them. Slowly but surely, as the century turned, the profligacy of the old aristocracy gave way to the respectability of the new middle class. Cynicism gave way to sententiousness, seduction to sentimentality – and the ostentatious and elitist playhouse to the accessible and affordable pamphlet, periodical and novel. It was Richardson, Fielding and Defoe who finally brought the word to the people in the form of the novel. And, more than any other, *the* words the eighteenth-century reader wanted to read were about how masters took advantage – especially sexual advantage – of their maids.

Daniel Defoe – son of a tallow-chandler, trader in wool and oysters, failed brick and tile maker, secret agent, editor, publisher, pamphleteer and 'father of the English novel' – came out with *The Life and Strange Surprizing Adventures of Robinson Crusoe* in 1719. There are no women on his deserted little island somewhere north of the Orinoco; and, in 24 years alone, Robinson Crusoe came to an antisocial conclusion. 'It was possible for me to be more happy in this forsaken solitary condition than it was probable I should ever have been in any other particular state in the world.'[46]

DeFoe's Moll Flanders lacked that option, but made the best of what she had. Having been left 'in bad hands' as a baby – her mother being convicted of petty theft and sent to the plantations to work – Moll was ordered into service at the age of 8. She begged off, managing to do needlework for her nurse in lieu of housework; but when the old nurse died after another six years, service became her only choice. Moll got a good place, being well qualified in three ways. 'First, I was apparently handsomer than any of them; secondly, I was better shaped; and thirdly, I sang better.' It was enough to get her courted, and seduced, by her mistress's two sons. The elder came first. Moll got 5 guineas for the first 15 minutes' foreplay, a handful of gold for going further, and 100 guineas for doing 'the last favour'. He promised marriage and talked about love; she 'was more confounded with the money than I was with the love', in particular with the gold, which 'I spent whole hours in looking upon'. Half a year later, Moll got an honest offer from the younger brother. The firstborn bought her off with a last £500, 'shifting off his whore into his brother's arms for a wife.' Thus was she made (for the time being) an honest woman.[47]

Moll Flanders was soon left a widow by the younger son; Pamela held out for, and married, an heir. Samuel Richardson – son of a London joiner, successful printer, publisher of the *Daily Gazeteer*, the *Daily Journal* and 26 volumes of *Journals* from the House of Commons – decided late in mid-life to put out a set of letters advising his daughter, then in service, on how to resist her master's advances. Pamela is seduced with presents ('he has given me a suit of my lady's clothes, and half a dozen of her shifts, and six fine handkerchiefs, and other of her cambric aprons, and four Holland ones'); she is kissed ('he talks of love without reserve; and makes nothing of allowing himself the liberty of kissing me'); she is sent away to a country estate, locked up, and kept under guard. Succour is hard to find. She appeals to a local farmer and his wife, but they are 'under obligations' to their landlord, Pamela's master; they 'expected repairs' from same; they 'were of the opinion that they ought not to intermeddle between a man of his rank and his servant.' She appeals to the minister of the parish, whose clergyman asks him to take her in; but the minister responds: 'What, and embroil myself with a man of Mr. B's power and fortune! Not I, I assure you. . . . It is what all young gentlemen will do.' She appeals to her procuress, who wants to help her escape, but admits she's been provided with a warrant from the local JP – who else but Mr B? – 'to get me apprehended, on suspicion of wronging him' in the event of her running away. Pamela even appeals to her master's sister who – like his labourers, clerks and domestic staff – sides with her brother. 'Confess the truth,' she says, 'that thou art an undone creature; hast been in bed with thy master; and art sorry for it.' Tell the truth, and everything will be all right. 'Some honest farmer may patch up thy shame, for the sake of the money.' In Richardson's version, the virgin wins in the end. But the sister-in-law gets in a last shot. 'You must not expect to have him all to yourself.'[48]

Though Pamela gets her man, Clarissa, Richardson's other heroine, does not. *Clarissa* is *Pamela*'s alter image: Pamela is a poor girl who holds out for – and snares – a rich husband; Clarissa is a rich girl who gets seduced – and wrecked – by a rake. Lovelace (the rake), his friends, his family and his enemies, are all explicit about his intent. Clarissa's loving aunt warns her niece: 'if his tenants had pretty daughters, they chose to keep them out of his sight'; this landlord 'kept no particular mistress, for he had heard *newelty*, that was the man's word, was everything with him.' To Clarissa's best friend's boyfriend, Lovelace 'was very happy, as he understood, in the esteem of the ladies' and 'pushed his good fortune as far as it would go'. And in Lovelace's own words, his wife – should he ever deign to take one – should be, among other things, a devoted procuress. She ought to 'be a *Lady Easy* to all my pleasures, and valuing those most, who most contributed to them; only sighing in private, that it was not *herself* at the time.' Like a good historian, Lovelace cites a few precedents: 'thus of old did the contending wives of the honest patriarchs; each recommending her handmaid to her lord, as she thought it would oblige him, and looking upon the genial product as her own.' Like a good philosopher, Lovelace elaborates: 'my predominant passion is *girl*, not *gold*; nor value I *this*, but as it helps me to *that*.' And like a good naturalist, Lovelace observes: 'Whatever our hearts are in, our heads will follow. Begin with spiders, with flies, with what we will, the girl is the centre of gravity, and we naturally tend to it.'[49]

The last of the three great eighteenth-century English fiction writers, Henry Fielding – playwright, barrister, magistrate, editor – published his great book *Tom Jones* in 1749. Its seductions are incidental. In the course of recovering his inheritance from Squire Allworthy (his unwitting uncle), and winning the fair Sophia Western (his neighbour's daughter), Tom has a few affairs. The first is with Molly Seagrim (Black George the gamekeeper's daughter). The effect is an 'alteration in her shape'. Molly doesn't much care; as she puts it to her mother and father: 'I shan't wash dishes for anybody. My gentleman will provide better for me.' She goes on, 'see what he hath given me this afternoon: he hath promised I shall never want money; and you shan't want money neither, if you will hold your tongue.' The question, as always, was how much. Tom deliberated at length. 'Her extreme poverty, and chiefly her egregious vanity . . . gave him some little hope.' He promised to provide for her 'beyond her utmost expectation', but ended up giving her the usual advice: 'She might soon find some man who would marry her, and who would make her much happier than she could be by leading a disreputable life with him.' Molly was accordingly pawned off on the philosopher, Square. Tom, by the way, lost no ground by all this with the Westerns. As the Squire put it to the parson: 'Ask Sophy there – You have not the worse opinion of a young fellow for getting a bastard, have you, girl? No, no, the women will like un the better for't.'[50]

Who was reading these things? Who were Defoe, Richardson and Fielding writing for? For the growing numbers of men – and especially women – of leisure who could afford to shell out two or three shillings for a novel. And for the tens of thousands of tradesmen, apprentices, schoolboys and *maids* who could afford to subscribe, at half a guinea a year, to one of the handful of circulating libraries set up in the mid-eighteenth century. This was Sam Johnson's 'nation of readers', Fielding's 'democracy, or rather a downright anarchy' of letters, when, in the words of Defoe, writing became 'a very considerable Branch of the English Commerce'.[51]

If, in the nineteenth century, Jane Austen, the Brontës, Charles Dickens, William Thackeray, Anthony Trollope, George Eliot, Henry James and Thomas Hardy had less to say about the sexual exploitation of the servant class, it probably was not because they, and their readers, did not care any more. It was because service as an institution was on its way out.

Adultery

Bastardy is bigger than bridal pregnancy. Fornication may show up in the parish registers; adultery will not. There are few hard, demographic or genetic, data on adultery, yet; most of the evidence is soft. It can be read, it can not be counted up.

Law

In the court records adultery, like fornication, was often motivated by gain. Sometimes, the other man's wife was paid off directly. Her fees could be surprisingly paltry. In seventeenth-century Somerset they included, on various occasions, 6*d.*, 1*s.*, 3*s.*, 5*s.*, 9*s.*, a bushel of wheat, 8 yards of canvas, a petticoat, or 50 faggots of wood. Marian Bradmore of Clevedon slept with three men for a blanket. Other times, adulterers paid the husband off. One husband 'much grieveth thereat but durst not speak thereof because he is indebted' to the lover of his wife.[52]

In the vast majority of trials for adultery, an offended husband sued an offending wife. In the minority of female plaintiff trials, the adultery is almost incidental – to impotence, disease or desertion. In, for instance, the collection called *Trials for Adultery: Or, the History of Divorces, Being Select Trials at Doctor's Commons*, wives sued husbands in just 9 out of 50 cases. Priscilla Adams sued for divorce in 1765, on the grounds that her husband was 'ruined' – and had, by the way, got another woman pregnant. Mary Willis, Elizabeth Bengough and Anne Parkyns had been abandoned for other women; Sarah Oliver sued Samuel Oliver for getting his wife's spinster sister with child. Dorothy Arnold sued John Arnold, who beat her too much: the 'blows, wounds, bruises, kicks, and other ill treatment' had reduced her 'to a very ill state of bodily health'; cohabitation had become a 'manifest danger of her life'. (She asked for, and was awarded, alimony at £20 per year.) Elizabeth Weld sued for impotence (case dismissed); and 15-year-old Dorothea Kingsman sued Ferdinando Tenducci for turning out to be a *castrato* (she got her divorce). Martha Robinson divorced Samuel Robinson for having 'frequently contracted the foul distemper, or venereal disease', and infected her twice. That – like being beaten or abandoned – was more than the double standard could bear.[53]

Most often, a husband sued for divorce on grounds of adultery by his wife. But frequently, divorce was not necessary. Many English husbands were more than happy to take cash. Colley Cibber, the Restoration playwright and England's Poet Laureate from 1730 to 1757, seems to have pimped for his wife, then sued her lover for insufficient compensation. Jane Phillips, who made the Cibbers' beds, swore under oath at the trial: 'he takes his money, lets him maintain his family, resigns his wife to him, and then comes to a court of justice, and to a jury of gentlemen, for reparation in damages.' Almost two hundred years later the Irish poet Caroline Norton asked Lord Melbourne, then Home Secretary and later Prime Minister, to get her poor husband George a job. George had lost his parliamentary seat in the last election, and was now living on her literary earnings. Lord Melbourne obliged Mr Norton, and struck up a close friendship with his wife. In 1836, Mr Norton had the gall to sue Melbourne for 'criminal conversation'; but as Norton conceded in court, he had often dropped his wife off at her lover's house. A few years later, Caroline's fellow suffragist Lady Emily Cecil sued for divorce from her own impecunious husband, Lord Westmeath. He had once suggested she advance the family fortunes by sleeping with her cousin, the Duke of Wellington. More particularly, his *mother* had tried to pimp her. 'Oh! My poor dear mother, she did not know what an old square-toes she was speaking to.' As Count Grammont put it at the Restoration court: 'in England it was thought astonishing for a man to feel jealousy about his wife' – at least, so long as he got a little money. But times would change. By 1886, Charles Dilke's political career came to a stop when he was accused by Donald Crawford, another Member of Parliament (MP), of adultery with his wife. Years before, Dilke had helped Crawford get placed in the Lord Advocate's office; evidently, that had not been enough.[54]

Letters, etc.

Sometimes, adultery like fornication was accomplished by force. Other times, carrots were used instead of sticks. This was the modus operandi of Samuel Pepys. When he took an unmarried woman to bed, Pepys tried to find her a husband. When he took a married woman to bed, Pepys found her husband a job.[55]

That is how it happened with Mrs Bagwell. In the thick of his early infatuation for Betty Lane, Pepys found himself attracted to a carpenter's wife. As the diary notes for 9 July 1663, 'thence I by water to Deptford, and there mustered the yard, purposely, God forgive me, to find out Bagwell, a carpenter, whose wife is a pretty woman, that I might have some occasion of knowing him and forcing her to come to the office again.' A week later, on 17 July, Pepys was 'saluted' on his way home by Bagwell and his wife '(the woman I have a kindness for).' They asked him in, and fed him some wine. Then on 9 August 'young Bagwell and his wife waylayd me to desire my favour about getting him a better ship, which I shall pretend to be willing to do for them, but my mind is to know his wife a little better.' Both ends were met. Mrs Bagwell was seduced on 15 November 1664; and Mr Bagwell got his post, as carpenter of the warship *Providence*, four and a half months later on 26 March.

Pepys had, like most moderately successful men, a wife – Elizabeth Marchant, a beautiful girl he jealously and passionately loved. Nor was his jealousy groundless. What Pepys spent 10 years (and probably the rest of his undiaried life) trying to do to lesser men, greater men were trying to do to him. Pepys was a man of humble beginnings – son of John Pepys, tailor, and washerwoman Margaret Knight. It was Sir Edward Mountagu, Pepys's first cousin once removed (Earl of Sandwich, Viscount Hinchingbrooke and Baron of St Neot's after the restoration of Charles II), who made his career. And it was Mountagu who tried to seduce his wife. One day, after he had tried the virtue of his own maid, Pepys was disconcerted to find

> my wife mightily troubled again, more than ever, and she tells me that it is from her examining the girle and getting a confession from her of all. . . . Reproaching me with my unkindness and perjury, I having denied my ever kissing her. As also with all her old kindness to me, and my ill-using of her from the beginning, and of the many temptations she hath refused out of faithfulness to me, whereof several she was particular in, and especially from my Lord Sandwich, by the solicitation of Captain Ferrers, and then afterward the courtship of [his son] my Lord Hinchingbrooke, even to the trouble of his lady. All which I did acknowledge and was troubled for, and wept.[56]

But that sort of indulgence was on its way out. Boswell, when he paid Rousseau that visit, wondered if he might be happier on the European continent than in Edinburgh or London. Why not play the gallant in Italy or France, 'where the husbands do not resent your making love to their wives?' Rousseau's answer was short: 'They are corpses. Do you want to be a corpse?' Boswell recanted at once. He knew 'he was right.'[57]

Literature

All this is borne out in novels. Defoe's Roxana, unlike Moll Flanders, is ruined, rather than made, by a husband. Married at 15 with a dowry of £2,000, Roxana is abandoned with five children not many years later. Then, having lost her money raising a legitimate family, she gets rich raising bastards by powerful men.

Her landlord is the first. As 'honesty is out of the Question, when Starving is the Case', she sleeps with him; she even procures her maid for him. And she is left with £10,000 sterling when he is killed on a highway in France. Consolation comes soon, in the form of a black box holding a banker's note for 2,000 livres a year. Its sender, her new lover, turns out to be an exceptionally good provider; 'his Person, and his Arguments, were irresistible.' This time she gives up not just sex, but sons. Great men, as Defoe points out, 'are always furnish'd to supply the Expence of their Out-of-the-Way Off-spring, by making little Assignments upon the Bank of *Lyons*, or the Town-House of *Paris*, and settling those Sums, to be receiv'd for the Maintenace of such Expence as they see Cause.' After a few years, well off and still young, Roxana goes back to England. She is seduced by a lord, but sets a snare for the king. If Defoe's hints are right, she gets what she wants. He has this to say about the English court: 'it is no Slander upon the Courtiers, *to say*, they were as wicked as any-body in reason cou'd desire them.' Roxana's friend set the standard. 'The KING had several Mistresses, who were prodigious fine, and there was a glorious Show on that Side indeed: If the Sovereign gave himself a Lose, it cou'd not be expected the rest of the Court shou'd be all Saints.'[58]

If Roxana gets rich by giving in to several seducers, Amelia barely scrapes by resisting one. *Amelia* is a classic story of adultery: rich man offers poor man a post, hoping to sleep with his wife. But Fielding does not like that plot. His real-life wife Charlotte, Amelia's model, has just died in his arms; he has not been well since *Tom Jones*; and he is a little soured, maybe, by what he has seen as a JP. In this, Fielding's last story, Amelia's husband Booth is deeply in debt. Booth's friend, Colonel James, suggests the usual way out. He has met a lord who 'loves the ladies' as well as Mark Antony, who 'if he once fixes his eye upon a woman, will stick at nothing to get her.' 'So what?', says Booth, being a little bit dense. 'Have not I shewn you,' James answers, 'where you may carry your goods to market?' Amelia's husband is shocked: 'What, to prostitute my wife!'

Having found a reluctant husband, the peer tries the woman herself. Amelia is plied with presents and compliments by 'one of my lord's pimps'. But her seduction is stopped by a confidante. As poor Mrs Bennett lets Amelia know, she has been bought off for the pox she got from the peer with a mere £150 a year – her lover having failed to get her husband the living he'd promised. This lord, she warns, is so addicted to 'novelty and resistance' that 'few of his numberless mistresses have ever received a second visit.' Amelia, having been put on her guard, resists; Booth pays his debts honesty, they raise a family, they stay chaste. And all live happily ever after – except the peer, who dies soon, a martyr 'to his amours, by which he was at last become so rotten, that he stunk above ground.'[59]

Defoe's and Fielding's complaints were anticipated on the Restoration stage. The subject of Restoration plots was seduction, and married women were seduced more often than maids. The titles give that away. There was *The Country Wife* (William Wycherley, 1674), *The Provoked Wife* (John Vanbrugh, 1697), *Madam Fickle* (Thomas Durfey, 1676) and *The London Cuckolds* (Edward Ravenscroft, 1682). Courtall (opposite Lady Cockwood in George Ethredge's 1668 *She Would if She*

Could), Horner (in *The Country Wife*), Woodall (in *The Kind Keeper*), Fainall (in William Congreve's 1700 *The Way of the World*) and Valentine (in Congreve's 1695 *Love for Love*) were leading men. The point of London play-going society was the point of the London play. Sometimes, the point was to find a maid. More often, the point was to find somebody else's wife. As Ravenscroft said in his *London Cuckolds* epilogue:

> Rouse up ye drowsy cuckolds of our isle,
> We see your aching hearts through your forced smile.
> Haste hence like bees, unto your City hives,
> And drive away the hornets from your wives.

As his Alderman Doodle whined: 'They'll never be content with our dull sport/ So long as Tories visit 'em from Court.' Or as Wycherley's Pinchwife put it to Horner: 'You have only squeezed my orange, I suppose, and given it me again; yet I must have a City patience' – that is, of course, a citizen's patience at being given horns by a peer. Toby, the man-about-town in Durfey's *Madam Fickle*, thought that in London 'a courtier can't walk the streets without being perpetually troubled in returning the compliments to some of his cuckolds: besides, they're so general a society here, that nobody minds 'em.' The provinces were not much different. In Salisbury, 'if a man is suspected to be a cuckold, he presently gets into office, either of constable, or head church-warden, that his degree may recover his disgrace.'[60]

But things were changing; the citizens were winning. By the turn of the century, the audience had changed. The cit was no longer in the pit; he was in the boxes, and the playwright was writing for him. Businessmen liked their lechers reformed. Loveless, for instance, in Colley Cibber's *Love's Last Shift* (1696), swears to his wife he will never leave: 'hitherto my soul has been enslaved to loose desires, to vain deluding follies, and shadows of substantial bliss: but now I wake with joy to find my rapture real.' The rake got a temporary reprieve in John Vanbrugh's satire, *The Relapse*, of a year later – here, Loveless cannot resist putting Berinthia, his good wife's pretty cousin, to bed. But the respite was short. The *Careless Husband* (Colley Cibber, 1705), a few years after that, was absolutely and forever reformed. Cibber's hero, Charles Easy, having been caught pretty much in the act, declaims to his long suffering wife: 'Oh! What a waste on thy neglected love, has my unthinking brain committed: but time and future thrift of tenderness shall yet repair all.'

> And like the ocean after ebb, shall move
> With constant force of due returning love.

As Woodall's man had put it in Dryden's *Kind Keeper*: 'Debauchery is upon its last legs in England.'[61] There were shades of that surprisingly early. The Stuarts had hardly got back the throne when Killigrew had two protagonists in *The Parson's Wife* (1663) complain about the king's *lack* of interest in women. These men did not like 'to see a king and queen good, husband and wife.' It made girls 'grow malicious' for want of attention; they 'call the king's chastity a neglect of them.' They missed the likes of Edward III or Henry VIII, 'no tame princes, but lions in the forest', kings who 'were properly called the fathers of their people, that were indeed akin to their nobility.' All this is more than a little ironic: Charles II was not a faithful husband. But the trend had already changed. 'For the City is a kind of tame beast.'[62]

Don Juan had had his day. The cit had had it with the aristocrat – not just in England, but all over the continent. The great story, *El Burlador de Sevilla*, was first put to paper by Tirso de Molina – a Spanish friar ordered in 1625 by a council of Castile to move out of Madrid and stop writing rude plays. *El Burlador* (variously translated as 'joker', 'trickster', 'seducer', 'rogue'), Don Juan Tenorio, was a rowdy noble in feudal Spain. He was nephew to Don Pedro Tenorio, the Spanish Ambassador to Naples; he was son of Don Diego Tenorio, Chamberlain to the King of Castile; and he was meant to be made Count of Lebria, a city 'subdued' by his father. He was handsome and well heeled ('The only bad day for me is when I'm short of cash. Everything else is nonsense'). And, in Molina's three acts, he managed to seduce several women. There was a duchess engaged to a Neapolitan duke, a peasant girl who saved his life when he got washed up on a beach, a village bride at her wedding ('nobility at a wedding is an ill wind that blows no good'). Juan's servant, Catalinón, did not like all that. 'I don't approve. We can't go on like this. One of these days you'll be tricked by your own tricks.' But Juan calmed him down. 'What are you afraid of? Isn't my father the Chief Justice, and isn't he the King's favourite?' Juan's days were, however, numbered. He made the mistake of asking a statue to dinner – the stone remains of a father he had killed defending his daughter's honour. The statue took hold of Juan's hand, and descended into hell. 'Don Juan,' he said, 'you are paying for the women you cheated.'[63]

So he was. Molina's Juan may have been nagged by his 'man', but he was done in – in the feudal tradition – by an injured noble. That was the old story. In the newer versions, Juan got outdone by Catalinón. In the popular performances that took up Molina's theme – in the travelling *commedia dell'arte* troupes, puppet shows, pantomimes and stage acts that played to plebeians all over England, Italy and France – Don Juan was upstaged by his valet. Leporello, Sganarelle, Passarino, Jacomo, Hans Pickering, Hans Wurst, Briguelle – all Catalinón incaranations – wanted (and get) more lines, more food, and (this is, after all, what *Don Juan* is about) more *girls*. They were tired of working, fighting and procuring for somebody else. They wanted – and, at long last, get – a piece of the action themselves.

They got it, especially, when the playwright was poor. That was most likely when Juan is a puppet, or in the *commedia dell'arte*. But it was true, too, of a few who wrote for the stage. Thomas Shadwell, for instance – the 'true-blew Protestant' who asked to be buried 'in Flannel with the least charge' – did not like Don Juan at all. His comedy, *The Libertine* of 1676, is a witless tirade against a malicious rake. It was very popular. Jacomo, Shadwell's Catalinón, gets the last word in this play; he reads the epilogue. A century and a half later, and much more to the point, another English playwright let his Catalinon get away with Juan's girl. William Moncrieff's *Giovanni in London or The Libertine Reclaimed* ('An Operatic Extravaganza in Two Acts') played to packed houses from 1817. This Don Giovanni (Juan) comes back from hell to find Leporello (Catalinón) has run off with Donna Anna, his old flame. She walks onstage in Act I, babe in arms. Giovanni is consoled, in the last act, with his own wife. But, true to nineteenth-century form, Juan wins her *because* he is a remade man. He swears 'eternal constancy and love' and adds, 'a milder path to find reform you've given/ than others did, since yours will lead to heaven.'[64]

A few continental blue bloods still honoured Juan's memory. Molière's *Don Juan ou le Festin de Pierre* did in 1665. Its author – son of the King's Upholsterer, patronised, pensioned and praised by Louis XIV – was reluctant to have his aristocrat condemned. This Juan, like old Juans, lacked remorse.

> What! You'd have me stick to the first Object that takes me, to renounce the World for that, and have no more Eyes for anybody? A fine thing indeed to pique oneself on a false Honour of being faithful, to bury oneself for ever in one Passion, and to be dead from one's very Youth to all other Beauties that may strike our Eyes. No, no, Constancy is fit for none but Fools. . . . All the pleasure of love lies in variety.

Even Sganarelle (Molière's Catalinón) is sorry, in the last scene, to see him go. And for the usual medieval reason – he will miss his pay. Juan gets swallowed up in hell while Sganarelle wails, 'My Wages! My Wages!'

Mozart too was sorry to see him go. His librettist, the great Signore Abbate Lorenzo da Ponte, was, among other things, expelled from seminary for reading subversive verses, prosecuted for public concubinage, and exiled from Venice for adultery. Mozart himself – patronised by the Archbishop Hieronymus von Colloredo in Salzburg and Vienna – annoyed people, grew poor, and was dead in 1791 at the age of 35. Together, they made Don Juan's most immortal incarnation, Don Giovanni. The opera opens with Leporello's (Catalinón's) chronic complaint: 'I must work day and night, endure the rain and the wind, eat badly and sleep worse, for a man who knows no gratitude. I'd like to act the gentleman myself, and give up the servant's life.' Leporello gets to masquerade, here, as husband to Elvira, Giovanni's virago of a wife. But Giovanni gets everybody else. Leporello has put their names in a book. There are quite a few of them:

> In Italy, six hundred and forty; in Germany, two hundred thirty-one. One hundred in France. In Turkey, ninety-one. But in Spain I've entered already a thousand and three! A thousand and three! You'll find, among them, country girls, waitresses, and city girls. We have countesses, duchesses, baronesses, princesses; women of every rank, of every shape, of every age. Blondes he praises for their gentleness; brunettes, for their constancy; grizzled heads, for their sweetness. In winter he likes them chubby; in summer, he likes them skinny. He calls the tall girl majestic and the short girl cuddly.

He seduces old women for the sheer pleasure of adding them to the list. His greatest appetite is for ripening virgins. But he doesn't care: let her be rich, let her be ugly, let her be beautiful. As long a she wears a skirt, you know what he'll do.[65]

The last, and least repentant, lover of *Don Juan* was Byron. Son to Captain 'Mad Jack' and great-nephew to 'wicked Lord' William, George Gordon came into his barony at 10, got a bastard daughter on his half-sister Augusta at 25, left England for Italy and got another bastard daughter on a woman of rank at 28, had affairs with uncertain numbers of poor women, and was frequently ill (of gonorrhoea) after Carnival. Byron's Don Juan does not go to hell; he does not even die, he just fades away. It is never clear, here, where Byron begins and Don Juan ends. Both were wiped out by the fever that stopped Byron's life, and his hand, in 1824 in the spring of his thirty-sixth year. Not before Juan was taken, in nearly 2000 flippant stanzas ('Hail muse! et cetera'), on a romp. This Juan is seduced: by a frustrated Spanish matron; by a stunning Greek girl on a beach; by a big, sultry woman in a Turkish harem; by Catherine the Great; by a good English wife. To Byron – who is writing feudalism's last gasps ('And where, oh where the devil are the rents?') – it is the last woman hunt.

> Oh thou *terterrima causa* of all *belli* –
> Thou gate of life and death – thou nondescript!
> Whence is our exit and our entrance. Well I
> May pause in pondering how all souls are dipt
> In thy perennial fountain. How man fell, I
> Know not, since knowledge saw her branches stript
> Of her first fruit; but how he falls and rises
> Since, thou hast settled beyond all surmises.
>
> Some call thee 'the worst cause of war', but I
> Maintain thou art the best, for after all
> From thee we come, to thee we go, and why
> To get at thee not batter down a wall
> Or waste a world, since no one can deny
> Thou dost replenish worlds both great and small?
> With or without thee all things at a stand
> Are or would be, thou sea of life's dry land!

Not that Juan/Byron does not have his regrets. Fidelity has its advantages ('how pleasant for the heart, as well as liver!') – and yet, 'Think you, if Laura had been Petrarch's wife,/he would have written sonnets all his life?/All tragedies are finished by a death,/all comedies are ended with a marriage.'[66]

The comedy of history was pretty much done. While Byron was amusing himself in Italy, Johann was unhappy in Germany. He was turning into Faust. The man of action's hands had finally been tied; and he'd been left with nothing to do but read. Goethe's hero, a great scholar, has mastered philosophy, medicine, theology and law. But he wakes up one night at his desk and whines: 'And here I am, for all my lore,/The wretched fool I was before.' He is powerless; he is out of cash; he is sick of his 'cave of sighs'. ('No dog would want to live longer this way!') He makes a pact with the devil, who feeds him a love potion; he seduces an innocent girl named Gretchen, then spears her brother in a street, leaves her mother to die of grief, abandons her infant to drown in a pond, and is saved anyway – for having tried? As Mephistopheles advises, 'stop playing with your melancholy,/that, like a vulture, ravages your breast,/the worst of company still cures this folly,/for you are human with the rest.' Nietzsche, a century later, was not impressed. As he put it in *The Wanderer and his Shadow*, in *Faust*: 'A little seamstress is seduced and made unhappy; a great scholar in all four branches of learning is the evildoer. Surely that could not have happened without super-natural interference? No, of course not! Without the aid of the incarnate devil the great scholar could never have accomplished this.' Maybe not.[67]

In two hundred years, Don Juan had come a long way – from a seventeenth-century devil-defying noble who seduced 1003 with ease, to a nineteenth-century intellectual who needed the devil's help to ruin one. Faust was nevertheless a man of action in the end. Don Juan since then, the twentieth-century Don Juan, has not left his 'cave of sighs' at all. He does not have sex with all women. He does not even have sex with one woman. He just *talks* about having sex, constantly. And he is usually very unhappy. Don Juan is the stuff of philosophy now. He is the stuff of literary criticism. That is so even when the philosopher writes plays – like George Bernard Shaw. Shaw's Don Juan is John Tanner, twentieth-century British socialist and MIRC (Member of the Idle Rich Class). Like his predecessors, Tanner descends into hell; unlike his predecessors, he has a long conversation with the devil. Hell here is 'a place where you have nothing to do but amuse yourself' and heaven is a haven for pure contemplation. Tanner does not sit well in hell: he is bored by the stupid and sordid 'Life Force'. He just wants to be left alone to think; only philosophers have ever been happy he is sure ('of all other sorts of men I declare myself tired'). The devil is disgusted. He says, in not so many words, 'one splendid body is worth the brains of a hundred dyspeptic, flatulent philosophers.' As Shaw says in the 'epistle dedicatory' prefixed to *Man and Superman*'s text, Don Juan's 'thousand and three affairs of gallantry, after becoming, at most, two immature intrigues, leading to sordid and prolonged complications and humiliations, have been discarded altogether as unworthy of his philosophic dignity and compromising to his newly acknowledged position as the founder of a school.' Most modern philosophers stick pretty much to that plot. For Kierkegaard, music was the apotheosis of life, Mozart the apotheosis of musicians, and *Don Giovanni* the apotheosis of Mozart ('Immortal Mozart! You to whom I owe everything – to whom I owe that I lost my mind'). But *Don Juan* was a period piece – he embodied the 'sensual' *idea* of his time. Camus thought Juan was never melancholy: 'every healthy creature tends to multiply himself.' These days, 'there is but one truly serious philosophical problem and that is suicide.' Foucault has Don Juan 'put into discourse'. So 'we shall leave it to psychoanalysts to speculate whether he was homosexual, narcissistic, or impotent.'[68]

Enough. Shaw and Camus, Molière and da Ponte were right. Don Juan is dead, and that is too bad for Don Juan. But the courtiers and philosophers were blind to the other side of the coin – the side that was clear to peasants and puppeteers. Juan's loss was Catalinón's/Sganarelle's/Leporello's gain. Once, big men in big houses had sex with most of the women and fathered most of the children. Now women, like money, are in wider circulation.

Kings

Kings set the standard. They had sex with maids – then saw to it that they were well married. And they had sex with matrons – then saw to it that their husbands were well employed. They lived in big houses, and they produced large numbers of bastards.

Some tallies, got by counting well-known mistresses and their well-known issue, suggest that the number of bastards got by English kings was not huge (Table 3.1). Henry I (1100–35) is the 'winner' in these inventories, with 20 'identifiable' and another 5 'possible' bastards. Charles II (1660–85) is second, with 14. William IV (1830–7) is next with 11; John (1199–1216) comes close, with 10. James II (1685–8) is credited with 6, George I (1714–27) with 4, George IV (1820–30) with 3 'acknowledged' and another 2 or 3 'unacknowledged' bastards. Other contributions, all small, make the kings on these lists – from William the Conqueror (1066–1100) to George VI (1936–52) – proud fathers of a total of 80 'identifiable' and another 29 'possible and doubtful' bastards.[69]

But the evidence that English kings – or imperial Roman, or medieval, or any other kings – made bastards is biased in at least three ways. First, kings, like other men, are more likely to acknowledge bastards born to high-ranking mothers, daughters of knights or better. Children by lower born women were apt to be forgotten. Second, kings, like other men, tended to acknowledge more bastards when they lacked legitimate heirs. That is, a recognised bastard was more likely to be in line for some share of the succession. And third, kings, like other men, were more likely to be maligned with 'lust' when they were otherwise maligned. An unpopular king was libelled, and sex was always one of the most popular slurs.

Bastards most likely to be counted were born to mothers who 'counted'. It was easier to acknowledge (harder not to acknowledge) children by mothers of rank. Arthur Lisle is a case in point. Edward IV, at age 19, met and seduced the Lady Elizabeth Lucy; her family had been Hampshire landowners since the fourteenth century at least. The upshot was Arthur, later Viscount Lisle, the last Plantagenet. Lisle's case applies, more or less, to most bastards on these lists.[70]

Bastards are more often counted when their fathers lack heirs. That gives illegitimate children – especially illegitimate sons – a better claim on their father's titles and lands. Henry I lost William, his wife's only son, in the wreck of the White Ship; he spent the rest of his reign marrying off bastard daughters to men who held lands he wanted to bind to his own, and promoting bastards' sons like Robert of Gloucester – who was, for a while, a contender for the throne.

Table 3.1 Bastards of English kings

King	Succeeded by a son?	'Identifiable' bastards	'Possible and doubtful' bastards	All bastards
William I	Yes	0	0	0
William II	No	0	1	1
Henry I	No	20	5	25
Stephen	No	1	4	5
Henry II	Yes	3	5	8
Richard I	No	1	0	1
John	Yes	7	3	10
Henry III	Yes	0	0	0
Edward I	Yes	0	1	1
Edward II	Yes	1	0	1
Edward III	Yes	3	1	4
Richard II	No	0	1	1
Henry IV	Yes	0	1	1
Henry V	Yes	0	0	0
Henry VI	No	0	0	0
Edward IV	Yes	3	1	4
Richard III	No	2	2	4
Henry VII	Yes	0	0	0
Henry VIII	Yes	1	0	1
James I	Yes	0	0	0
Charles I	Yes	0	0	0
Charles II	No	14	0	14
James II	No	6	0	6
George I	Yes	4	0	4
George II	Yes	0	0	0
George III	Yes	0	0	0
George IV	No	3	3	6
William IV	No	11	0	11
Edward VII	Yes	0	1	1
George V	Yes	0	0	0
George VI	No	0	0	0

Note: Edward VI is omitted because he died at 15, too young to be sexually libelled or to have fathered an heir. Edward VIII, who abdicated in 1936, is also omitted. Neither was ever crowned.

This pattern, too, repeats itself.[71] The claim that heirless kings counted more bastards can be statistically made. Altogether, the 19 kings who *were* succeeded by a son are credited with just 22 'identifiable' bastards; the 12 kings who *were not* succeeded by a son left at least 58. When 'possible and doubtful' bastards are added in, kings with a secure succession left 35 illegitimate children; kings not succeeded by a son left 74. For 'identifiable' bastards, the difference in means is 1.16 (for kings succeeded by sons) versus 4.83 (for kings not succeeded by sons); for all bastards – 'identifiable', 'possible', plus 'doubtful' – the means are 1.84 versus 6.17. These are significant differences. As bastard counters themselves point out, 'it is one of those perverse ironies of our history that neither

Table 3.2 The reputations of Norman to Tudor kings

	Murdered or deposed	*Not murdered or deposed*
More promiscuous	Richard II	William II
	Edward IV	John
	Richard III	
Less promiscuous	Edward II	William I
	Henry VI	Henry I
		Stephen
		Henry II
		Richard I
		Henry III
		Edward I
		Edward III
		Henry IV
		Henry V
		Henry VII
		Henry VIII

Note: Edward VI is omitted as he was never crowned.

Henry I, who fathered more bastards than any other English king, nor Charles II, who ran him a close second, was able to pass his crown to a legitimate son.' Not, as it turns out, ironic at all.[72]

The third claim, that a weak king is a maligned king, can be statistically made as well. The weakest of kings is deposed, and a deposed king is almost always despised. Of the 19 men who ruled England from William I to Henry VIII – that is, of the medieval Norman to Tudor kings – 5 were deposed or killed (Table 3.2). At least 3 of these were seriously sexually libelled: Richard II, Edward IV and Richard III. John, also lecherous, was not deposed, but came close: he conceded *Magna Carta* to his barons, and England to Innocent III. And William II, extremely sexy, was brought down by an arrow in the New Forest that *might* not have been innocuous – it could have been fired with a little help from his little brother. Other exceptions are as easily explained. Edward II was done in by his wife, for the sake of his son: that gave posterity less cause to call him promiscuous. And Henry VI is not besmirched, though Edward IV – his usurper – is. That probably has something to do with the fact that, in the end, Henry VI's Lancastrian line won the throne, and posterity was kind. In short, a king with legitimate sons lacked a motive to expose his bastards; and everybody else lacked a motive to expose his bad habits.

The remaining 12 of the first 19 English kings – kings who were never killed or deposed – are less likely to be criticised as sexy, and more likely to be admired for their 'chastity'. They are: William I ('chaste' enough to make some think him impotent – till he got 10 children on his wife), Stephen (who had the grace to give his kingdom away to a stronger man), Henry II (conceded an occasional affair), Richard I (lately supposed by some to have been less interested in women than men), Henry III (also conceded an occasional affair), Edward I (seldom slandered), Edward III (unrivalled and unlibelled) and Henries IV, V, VII and VIII (Tudor kings and their Lancastrian ancestors, all well regarded by Tudor chroniclers). Even strong Henry I, who made a career of advancing his bastards, was more often defended than maligned. To William of Malmesbury – who was patronised by Henry I's best known bastard – 'he was led by female blandishments, not for the gratification of incontinency, but for the sake of issue; nor condescended to casual intercourse, unless where it might produce that effect.'

Overall, then, 3 out of 5 killed or deposed Norman to Tudor kings were excessively sexually libelled; while 12 of 14 undeposed kings were called 'impotent' or 'chaste', or more or less ignored. This is a strong trend. A weak king is a maligned king; and the most malicious of slurs is a sexual slur – in England as everywhere else.[73]

Evidence on the production of bastards by England's kings is biased, then, toward unpopular kings, and toward kings without legitimate sons. But the production of royal bastards shows another real bias. More recent kings – kings since around the turn of the seventeenth century – have probably had fewer opportunities. As men in England, then on the European continent, in North America and finally worldwide lost power, they gradually lost access to the opposite sex. This is the point I am trying to make. It is made, partly, by figures on household size. And it is made partly by words.

On 14 October 1066, at Hastings, England was conquered by William the Bastard.[74] He had been conceived in the usual way. One day in 1027, in the little French town of Falaise, Robert I, sixth Duke of Normandy, had got 'so smitten . . . as to form a connection' with a tanner's daughter. Nine months later William, her bastard son, was born. The girl, Herleve, found her fortunes immediately advanced. Her father Fulbert obtained an office at the ducal court; her brothers Osbert and Walter started to show up as witnesses to important government documents; and Herleve herself, shortly after her bastard's birth, was married off to a Conterville vicomte. She gave her new husband two legitimate sons; they became Robert, Count of Mortmain, and Odo, Bishop of Bayeux. This is the type case. It came to light when William the Bastard succeeded his father at the age of 7, in 1035.

Most bastards, unlike William, probably do not come to light. None of William the Bastard's own bastards has. On the other hand, unlike his father, William lacked a motive to expose them. He married Matilda – granddaughter of King Robert II of France, niece of King Henry I, and daughter of Baldwin V, Count of Flanders, King Philip I's protector. Then, between 1050 or 1051 (when they married) to 1083 (when she died), Matilda bore four legitimate sons. Two lived to succeed William as kings of England.

William I, a paradigm of strength, seems to have been a paragon of discretion. He was, to an English monk who had once lived at his court, a greedy man (he was 'so very stark, and deprived his subjects of many a mark') and mean (men 'had to follow out the king's will entirely if they wished to live or hold their land'). Everybody knew he loved food: when he died, they stuffed him in a sarcophagus that was too narrow and short, and 'his bowels, nourished with so many delicacies, shamefully burst, revealing to wise and foolish alike how vain is the glory of the flesh.' But they never called him a libertine. To the Norman monk at Jumièges who dedicated his book to the Conqueror, he was 'temperate' and pious: 'whenever his health permitted' he went to mass. To William of Poitiers, the Bastard's Norman chaplain, 'everywhere people greeted him with joyous applause and sweet songs' because 'nothing impure was beautiful' to this king. That verdict held up. William of Malmesbury – the twelfth-century abbey librarian who dedicated his *Historia Novella* to Robert of Gloucester, William's bastard grandson by Henry I – says that 'in addition to his other virtues he, more especially in early youth, was observant of chastity.' And was often thought impotent, in fact – till he obtained several heirs from his wife. Nevertheless, William I 'conducted himself, during many years, in such wise, as never to be suspected of any criminal intercourse.' Others may say otherwise; 'I esteem it folly to believe this of so great a king.'[75]

Unlike his father, William II never married. And unlike his father, William the Red ('Rufus') was supposed to be promiscuous. The pattern, in short, was set: William I, a strong husband with 10 legitimate children, was 'chaste'; William II, a bachelor without heirs who was killed by a New Forest arrow, was not. In particular, to Anselm – the Archbishop of Canterbury with whom Rufus ran in – William II was reprehensibly loose. To Eadmer, who was close to Anselm, incest and sodomy were common; besides, young men wore their hair too long at court. Orderic Vitalis – the twelfth-century Norman monk whose account followed Anselm's – says they shaved their heads in front ('like thieves') but let it grow long and curled it with irons in back; their robes dragged on the floor and their sleeves were too long ('impeded by these frivolities they are almost incapable of walking'); they wore shoes with pointed toes ('excrescences like serpents' tails'); they grew little beards ('they revel in filthy lusts like stinking goats'); and they minced. William of Malmesbury says 'they unwillingly remained what nature had made them; the assailers of others' chastity, prodigal of their own'; besides which, 'troops of pathics, and droves of harlots, followed the court'. Orderic calls Rufus 'far too prone to pride, lust and other vices'; 'wanton and lascivious'; a sodomite. But they liked girls, too. 'Our wanton youth is sunk in effeminacy, and courtiers, fawning, seek the favours of women with every kind of lewdness' When William the Red died, no alms were saved for the poor. But 'soldiers, lechers, and common harlots lost their wages through the death of the lascivious king.'

As they lived at home, they lived abroad: Rufus's men raped the countryside. Like other royal medieval households, William II's was often itinerant; and, when they travelled, they took what they needed where they found it. Eadmer says they 'made a practice of plundering and destroying everything'; they 'laid waste all the territory through which the King passed.' What they could not consume on the spot they packed up to sell at market, or burned, or raped. 'What cruelties they inflicted on the fathers of families, what indecencies on their wives and daughters, it is shocking to think of.' When village people heard the king was on his way, they hid in the woods. Henry I, Malmesbury says, found it necessary to issue an edict restraining 'the exactions of the courtiers, thefts, rapine, and the violation of women'; guilty parties would be blinded or deprived of their 'manhood'.[76]

Had Rufus lived past his early forties, had he not been cut down by a stray arrow in the woods one day in his prime, he might have got legitimate heirs on a legitimate wife, or made posterity sure of some of his bastards' names. But William the Red died, in pursuit of red deer, on 2 August 1100; Henry I succeeded three days later. Within four months, 'not wishing to wallow in lasciviousness like any horse or mule' (Orderic Vitalis), Henry took a wife. On her he fathered two legitimate children, but had the bad luck to lose the boy at sea in 1120. So he spent the last 15 years of his life and reign trying to secure his kingdom. He tried, first, a second marriage that produced no issue; he tried, second, owning up to and marrying off 20 or more bastard daughters and sons. One Robert was made Earl of Gloucester in 1121 or 1122, and married off to the Lord of Glamorgan's heiress; Rainald was made Earl of Cornwall in 1141; another Robert got married off to an Avranches heiress; William held several knight's fees and witnessed several charters; and several more sons died young. At least two bastards ended up in church: a Fulk ('*Fulcone filio regis*' in an abbey history) was probably an Abingdon monk; a Maud (called sister of the Queen of Scotland in a chronicle) was Abbess of Montivilliers. Many of Henry's other daughters got married – to the Count of Perche, the Lord of Montmirail, the Lord of Montmorency, the Lord of Breteuil, the Duke of Brittany, the Vicomte of Maine, the Earl of Galloway, the King of Scotland: men whose lands he intended to bind to his own. The mothers of Henry's children came from all over Britain. Ansfride was the widow of a knight on the Abingdon abbey estates; Sibyl was Robert Corbet of Alcester's daughter; Edith was the Lord of Greystoke's daughter; Nest was the daughter of a prince of South Wales; Isabel was a daughter of the first Earl of Leicester, and a grand-daughter of Hugh of Vermandois. Henry found some of these women husbands. Sibyl got Herbert FitzHerbert, the son of a chamberlain; Edith got married off to a royal constable, Robert de Oilli; Nest got Gerald of Windsor (their grandson was Gerald of Wales); and Isabel ended up with Gilbert FitzGilbert, the first Earl of Pembroke.

In spite of all that, Henry I was not often maligned. Orderic found it enough to note that, 'possessing an abundance of wealth and luxuries, he gave way too easily to the sin of lust; from boyhood until old age he was sinfully enslaved by this vice.' Malmesbury is not just terse, he is apologetic. He insisted that Henry 'was free, during his whole life, from impure desires;' he was 'master of his inclinations, not the passive slave of lust.' To an archdeacon of Huntington, Henry I had three virtues (he was wise, he was warlike, he was rich) and three vices (greed, cruelty and 'wantonness, for, like Solomon, he was perpetually enslaved by female seductions'). To Robert of Lewes – the twelfth-century Bishop of Bath who prob-ably wrote the *Gesta Stephani* – this Henry was 'the peace of his country and the father of his people.'[77]

When he died, for all his efforts heirless, Henry was succeeded by Stephen, his sister Adela's son. That succession was disputed throughout his reign – mainly by Matilda, Henry I's one surviving legitimate offspring, and by Robert of Gloucester, her bastard half-brother. Stephen was, as an effect of this contest, weak; and his contemporaries were not nice. 'He was soft and easy-going and did no justice'

(*Anglo-Saxon Chronicle*); 'for the solemnities of the court, and the splendours of the royal household, which had come down from earlier times, now entirely vanished; there was peace in no land' (Robert de Torigni); England was 'a haunt of strife, a training ground of disorder, and a teacher of every kind of rebellion' (*Gesta Stephani*). The funny thing about Stephen is that, unlike other kings whose kingdom came under assault, his virtue did not. Almost nothing, good or bad, was said about his private life. His contemporaries found him, among other things, 'kindly and gentle' – perhaps because his kingdom was so easily given up.[78]

In the end, it went to Matilda's son. On 18 May 1152 Henry of Anjou married Eleanor, heiress of Aquitaine, and started an empire. A little over a year later, by a treaty signed 6 November 1153 at Winchester, Stephen made his second cousin Henry II his heir – in preference to William, his still surviving son. Less than a year later, Stephen died; and, at his leisure, Henry II – just 21 – was crowned on 19 December. Henry went on to have several legitimate sons. Two lived to succeed him as kings of England. But there were other women, and other children. Henry II also had two well-known, well-placed, bastards. One was Geoffrey 'Plantagenet', eventually bishop of Lincoln, chancellor, and archbishop of York; the other was William 'Longsword', who became Earl of Salisbury. There are other 'doubtful' bastards, though Henry's reputation for 'concupiscence' is not. As Eleanor's sons grew up, they bid for power; and their mother was too strong a backer. After the siege of Rouen in 1173, Henry had Eleanor locked up. In the mean time, he amused himself. He is supposed to have coveted the sister of Roger of Clare, Earl of Hertford – the 'most beautiful woman in the realm'. A little later Eudo de Porhoet, who had been Brittany's count, complained he held his daughter as hostage in 1168 and got her pregnant 'treacherously, adulterously, and incestuously'. He probably fathered another celibate bastard – Morgan, a provost of Berkeley and bishop-elect of Durham, who went to Rome in 1213 to get papal confirmation, but decided he could not deny his spurious descent from Henry – on Nest, Sir Ralph Bloet's wife. And there were rumours he had debauched his daughter-in-law to-be. Alice, the sister of Philip Augustus of France, had been meant to marry Richard, Henry's son; and when Philip asked why he did not want her, Richard answered that he had 'on no account whatever take his sister to wife; inasmuch as the king of England, his own father, had been intimate with her, and had a son by her.' Witnesses were produced 'to establish that fact'. There were humbler girlfriends. There was 'fair Rosamond' Clifford – whose tomb at a Godstow nunnery Henry had draped in silk cloths; there was 'Bellebelle', provided with cloaks and other clothes on an 1184 pipe roll. Others did not have names. Gerald of Wales – a clerk at court, who Henry would not make bishop of St David's because Gerald was 'bound so closely by ties of blood to the magnates of Wales' – said the king 'became an open violator of the marriage bond.' His vassals, says Ralph Niger – a friend of Becket's – hid their daughters and wives when the king was in other towns, since Henry 'was a corrupter of chastity, and followed his father in committing crimes.' Life was, as always, grand at court. Walter Map – who wrote his *Courtier's Trifles* as one of Henry's clerks – says the king was 'sober, modest, pious, trustworthy and careful, generous and successful', but 'troubled by erotic dreams' that he could not stop. Even Becket, who imitated his king, kept hundreds of retainers, servants, guests – and, as chancellor, wolves and monkeys, 'harlots and lechers' at his court. Becket's friends said he did not notice: that he avoided sex, even when his doctors advised it. Nobody said that about Henry. It was after all Ranulf de Broc – keeper of the king's whores – who looked after revenues from Canterbury when Becket lived in France; and it was de Broc who led the siege when Becket was murdered in the cathedral.[79]

Richard I Lionheart spent just six months of his ten-year reign in England. The rest of the time, he was in his mother's native Aquitaine or away on crusade. As Roger of Howden – professor of theology at Oxford, possibly chaplain to Henry II – put it in his thirteenth-century *Gesta Regis Henrici Secundi*, on the continent Richard 'carried off his subjects' wives, daughters and kinswomen by force and made them his concubines' and when he was done 'handed them down for his soldiers to enjoy.' Further east, there were more of the same: girls, e.g. of Acre, made themselves available to the king and other crusaders. As Imad ad-Din, secretary to Saladin, pointed out: 'tinted and painted, desirable and appetizing, bold and ardent . . . they offered their wares for enjoyment . . . interwove leg with leg, caught lizard after lizard in their holes, guided pens to inkwells, torrents to the valley bottom, swords to scabbards, firewood to stoves.' Richard was, Roger of Howden was sure, 'sensible of the filthiness of his life.' But contrite. One day in 1195, being advised by a hermit to be 'mindful of the destruction of Sodom', he got sick, confessed 'the guiltiness of his life', and resolved to live again with his wife 'putting away all illicit intercourse'. Back to Berengaria of Navarre he may have gone; but no heirs ensued. The reference to Sodom (and Richard's close ties to Philip Augustus of France, and his apparent lack of interest in his wife) has been enough to make some suspect *coeur de lion* avoided sexual contact with women. That, given the sources, seems unlikely. But bisexuality is a possibility. Sexual contact – with women *or* men – had cemented aristocratic ties since ancient Greece and Rome at least. John Gillingham, Richard's modern biographer, writes: 'what Richard and Philip were doing was not making love but making a political gesture.' They may have been doing both. In England William II, Richard I, Edward II and James I, at least, have been suspected of sexual contact with men. Probably rightly; but probably not exclusively. Even as he lay dying, if Walter of Guisborough – who wrote in the early fourteenth century – is right, Richard was provided with whores. 'He was strongly lubricious and burning with love for women, unwilling to curb his pleasures, belittling safe counsels, not worrying about his wound.'[80]

Being heirless, Richard left the kingdom to his youngest brother, his father's favourite, John. There were losses to Philip Augustus, and concessions (e.g. *Magna Carta*) to English barons. John ('Lackland' and 'Softsword') was a weak, maligned king. He was uxorious – very fond of his wife, according to Roger of Wendover, the early-thirteenth-century St Albans monk. And, according to Matthew Paris – Wendover's successor, the great thirteenth-century St Albans chronicler – he was fond of other men's women. John had some trouble raising an heir. He got none on his first wife, Hadwiga of Gloucester, and divorced her soon after being crowned in May 1199. He had better luck with Isabella of Angoulême, the 12-year-old wife he took just three months later, and on whom he eventually got five legitimate children. But, in the mean time, John made his magnates mad. According to more than one source, he 'seduced their more attractive daughters and wives'. Among the king's known women were: Suzanne (listed *domicella, amica domini Regis* in a Misae roll); 'queen' Clementia (named by a Tewkesbury monk); Hawise (the widowed countess of Aumale whose son, says a pipe roll, was pardoned his relief payment on behalf of his mother's efforts); another Hawise (who, in a memoranda roll entry, was ordered a pension of a penny a day); 'Alpesia, the queen's damoiselle' brought under escort to the king on 14 July 1214; and the wife of Hugh Neville, chief forester, who according to a Christmas 1204 entry on an oblate roll promised the king 200 chickens that she might lie one night *with her husband*. The upshots were bastards – several of whose names are known. One, John, was supported by the see of Lincoln; Henry FitzRoy was given lands in Cornwall in 1215; a daughter, Joan, was married to a Welsh prince, Llywelyn; Richard captained troops in a baronial revolt; Geoffrey emerged from obscurity to command a troop of mercenaries in 1193; Osbert Giffard's fate is not clear.[81]

By his wife Eleanor of Provence, Henry III had five legitimate children. Two were sons, and one – Edward 'Longshanks' – succeeded him. Henry, in short, lacked a motive to bring his bastards' names to light. And, for most of his reign, people lacked a motive to malign him. There were a few exceptions. Just before he married, in 1236, the Earl of Kent spread rumours that Henry was 'squint-eyed, silly, and impotent'; that he was 'more a woman than a man'; that he was 'entirely incapable of enjoying the embraces of any noble lady'. Henry and Eleanor proved him wrong. A little later, in 1243, Matthew Paris found the king 'passing the winter at Bordeaux and wantonly staying there' – the Countess of Biard extorting money (up to £30 sterling) daily. He might have had other chances abroad. When this Henry fought off his barons at Northampton and Lewes, political songs called his peers chaste, but said there were 700 'stinking bawds' and 'so many strumpets' in the royalist camp. And he might have chances at home. Henry de la Mare and his family had the job of 'looking after the harlots who follow the king's court'. Toward the end of his reign, having won his last battle at Evesham, Henry 'gathered together an even larger household than before'. People had complained about that. In 1258, after he had lost to his brother-in-law at Lewes, the *Provisions of Oxford* made a note 'to amend the households of the king and queen.'[82]

In 1272 Edward I got the throne, uncontested, from his father, then ruled England for 35 years with a firm hand. In 1253 he made Eleanor of Castile his wife; they produced 14 children at least. In 1299, having lost Eleanor in 1290, he married 17-year-old Margaret, half-sister to Philip IV of France; she gave him another daughter, and two more sons. By these several children, there were several grandchildren. The messengers who told Edward his daughter Elizabeth's sons had been born got paid £26 and £30 for making him happy; when his daughter Margaret's son was born, the messenger got £126. There was never, in short, any question that Edward I was well supplied for succession. And there was never much doubt that the winner of wars in Wales and Scotland ('Arthur' – whose remains Edward had dug up at Glastonbury in 1278 – 'had never the fiefs so fully') had a fair reputation. Like his father, Edward had no motive to remember his bastards; and most contemporaries lacked a motive to malign him. After his father lost a battle with his barons, the *Song of Lewes* said Edward changed like a leopard: 'wrong gives him pleasure and is called right; whatever he likes he says is lawful'; and at the end of his father's tenure, he is supposed to have quarrelled with the Earl of Gloucester on account of the 'excessive intimacy which Edward was said to have with the earl's wife.' Then they made him a king, and the rumours stopped. Opportunities probably did not. 'Longshanks' was reluctant to walk through Oxford's west gate 'on the pretext of the prayers of the sainted virgin Fritheswyte' – an Anglo-Saxon who had been chased down on the spot by a Mercian prince. And at Cambridgeshire, Johannes of Windreshull had custody of a manor for finding somebody 'to keep the whores in the king's army'. Edward I's court was, at any rate, rich. His first Statute of Westminster of 1275 promised to do away with prises: agents who took things for the king or his castles, then kept cash from their creditors, would be made to pay from their own lands and chattels, or 'be in prison at the king's pleasure'. They bought a lot in. *Records of the Wardrobe and Household* of 1285–9 list animals (bears, boars, leopards, lions, stags, steers and so on), fancy clothes (multicoloured cloths, muslin, samite, silk, feathers, gold fringes, ermine and miniver furs), fun and games (minstrels and musicians, dancing and gaming), and food. They ate 1,500 cattle, 3,000 sheep, 1,200 pigs and 400 bacon carcasses in six months; they had 1,742 chickens, 22 pheasants, 17 dozen partridges and 16 dozen mallards at Christmas.[83]

On 19 June 1299 Edward II, at the age of 15, got engaged to 8-year-old Isabelle, daughter of Philip IV 'the Fair' of France – on the same day his father got engaged to Philip's half-sister, Margaret. Thus were the seeds of his destruction sown: he would in the end be outmanned by his powerful stepmother, by his powerful wife, and by his own powerful sons. By 1321, Isabelle had given her husband a pair of girls and a pair of boys. Half a dozen years later, on 13 January 1327 at Guildhall in London, 'the queen reconquered the realm of England for her eldest son.' Six articles of accusation were read, and Edward II stepped down. They said the king was 'incompetent'; he was given to 'unseemly works and occupations' (he liked boats); he had disinherited his nobles and destroyed his church; he was not just; he had lost wars in Gascony, Ireland and Scotland; he had 'stripped his realm' and ruined his kingdom. They killed him 'with a hoote broche putte thro the secret place posterialle' at Berkeley Castle soon after. Unlike his father, and unlike his son, this king was weak and maligned. Some people said he was fond of women. John of Warblinton succeeded his father Thomas at looking after 'the whores of the king's court' in around 1317; there was at least one bastard, Adam, who fought along with his father in Scotland. And some people said he was fond of men. The usual father–son conflict was, between Edward II and his father, fierce. Edward I is said to have called his boy a 'baseborn whoreson', to have torn out his hair, knocked him about and kicked him, and threatened that 'if it were not for fear of breaking up the kingdom' England should never be his. In 1307, the king threw Piers Gaveston out of the country 'for certain reasons': he was 'the most intimate and highly favoured member' of the prince's chamber. Piers would be brought back to England when Edward II became king; and he would be beheaded by his barons five years into his reign. The lords, and the commons, had got the better of Caernarfon a year before. The 41 *New Ordinances* of 1311 asked the king to stop taking corn, merchandise and other kinds of goods without paying for them, and 'to live of his own without taking prises other than the ancient, due and rightful ones.' People who took things for the king without paying were chased down 'with hue and cry', tried as 'a robber or thief', and if convicted hanged. Seven years later, the *Household Ordinance* pared the king's staff down. Prostitutes, in particular, were to be locked up after a third offence; and men were asked not to keep a wife 'or other woman' at court.[84]

They made Edward III king in 1327, at the age of 14. To Froissart, a Hainaulter like this Edward's wife, 'his like had not been seen since the days of king Arthur', whose round table he copied – though Jean le Bel, a Hainaulter whose chronicle Froissart reworked, had him raping the countess of Salisbury after a game of chess. ('She was one of the most beautiful young ladies in the land.') Philippa of Hainault gave her husband plenty of children: there were at least four daughters and five sons. But Philippa died in 1369; and, after many wins, the third Edward's reversals abroad meant reversals at home. Having lost battles in France, the Good Parliament of 1376 had Alice Perrers and a few ministers impeached: 'she makes wantons of wives and widows. . . . I swear to God you won't find a greater bawd between Heaven and hell', complained William Langland's *Piers Plowman*; and the *Brut*, England's most popular chronicle, was sorry to say the king had 'lechery and moving of his flesh haunting him in his old age, wherefore rather, as it was to suppose, for unmeasurable fulfilling of his lust, his life shortened the sooner.' Edward died a few months after his Good Parliament met, and Ms Perrers's bastards suffered as a result. Alice's girls, Joan and Jane, married unspectacularly – to a Kingston-upon-Thames lawyer called Robert Skerne, and to a man called Richard Northland, otherwise unknown. John, Alice's 'valiant bastard' boy, led a mutiny in Lisbon against one of his many legitimate half-brothers – the Duke of York, Edmund Langley. In the good days, Edward had spent pretty freely. He had had Windsor Castle rebuilt at a cost of £51,000 at least; he had wanted it to be 'the Versailles of the age'. Like other kings, he got commodities at good rates. The *De Speculo Regis*, an anonymous tract, complained the king's agents paid 3*d.* for 5*d.* worth of hay, 3*d.* for 8*d.* worth of barley, 3*d.* for 1*s.* worth of beans. 'Where is the justice in this? It's not justice, but rape.' Every Parliament from 1343 to 1355 petitioned against purveyance; the great statute of purveyors passed in 1362 spelled out how 'buyers' should come up with fair prices. 'Because no free man ought to be assessed or taxed without the common consent of parliament.'[85]

Like his prolific grandfather Edward I, Edward III left too many legitimate sons behind. Richard II – firstborn son of his late firstborn son – succeeded in 1377 at the tender age of 11, then fought off his uncles for 22 years. In the autumn of 1399, Richard got locked in the Tower of London; on 30 September, 33 articles of deposition were read against him. They alleged the king had said (no. 16) that the laws of the realm were 'in his mouth' or 'in his breast', and (no. 26) that the lives and goods of his subjects 'were his, and subject to his will.' They got Richard to 'confess, acknowledge, recognize, and from my own certain knowledge truly admit that I have been and am entirely inadequate and unequal to the task of ruling and governing' England; and they put Henry Bolingbroke, John of Gaunt's son, on the throne. Richard II had married Anne of Bohemia in 1382, but got no issue by her; Bolingbroke had cut Richard's second marriage to 7-year-old Isabelle of Valois short. As of the day he lost power just three years later, none of Richard's bastards' names were well known. He had chances, though. The taxes that made his peasants revolt (tax men liked to lift girls' skirts 'to test whether they were corrupted by intercourse with men' and got their parents to pay 'rather than have them touched in such a disgraceful way') fed lots of people at court. They hung out at new bathhouses built at Eltham and Sheen – paved with 2,000 painted tiles; they were entertained by John Gower – who wrote a love poem on a royal barge at the king's request, and by Geoffrey Chaucer – whose *Tale of Melibee* might have been written for Richard. In all, his household cost as much as £53,200 a year; and his parliaments asked him, again and again, to size it down. On the last day of 1387, five appellants – Bolingbroke and four friends – did an inventory at Westminster: there were 100 in the buttery alone, and 'superfluous numbers' on the kitchen and other staffs, 'whose excess they pruned'. Then, in January 1397, the commons singled out the expense of keeping so many *ladies* at court. 'As to the fourth article, touching the expense of the royal household and the residence of bishops and ladies in his company, the king was greatly aggrieved and offended.' This was too much even for the lords, who said any man guilty of such criticism was guilty of treason. But, if the chroniclers are right, the commons were mad for a reason. Thomas Walsingham – another monk from St Albans – called the king's men 'knights of Venus rather than of Bellona: more effective in bed than in the field'; they beat, wounded, killed and robbed men, 'raped and ravished both married and unmarried women.' And a monk from Evesham said the king burned the candle at both ends, 'sometimes staying up half the night, other times right through till morning, drinking and indulging himself in other unmentionable ways.'[86]

The wars of the roses followed. The house of Lancaster (red rose: Edward III's third son, John of Gaunt's line) contended with the house of York (white rose: Edward III's fourth son, Edmund Langley's line) for most of the hundred years England and France were at war. To Edward Hall (who dedicated his chronicle to Edward VI, Henry VIII's one legitimate son) and to Raphael Holinshed (who lived in and glorified the reign of Henry VIII's daughter, Elizabeth I), Lancastrian kings were good, chaste kings. Yorkist kings were not. Henry IV, Bolingbroke, 'reigned thirteene yeares, five moneths and od daies, in great perplexitie and little pleasure' (Holinshed). To Shakespeare, who followed Holinshed, Bolingbroke's son Henry V sowed oats as a boy:

FALSTAFF: What time of day is it, lad?
PRINCE OF WALES: What a devil hast thou to do with the time of the day? unless hours were cups of sack, and mintes capons, and clocks the tongues of bawds, and dials the signs of leaping-houses, and the blessed sun himself a fair hot wench in flame-colour'd taffeta, I see no reason why thou shouldst be so superfluous to demand the time of day.

But, being a good Lancastrian king, he reformed. Henry V 'beyng a prince of honor, a prince of youth, a prince of riches, did continally abstain from lasciuious liuyng' (Hall); 'this Henrie was a king, of a life without spot' (Holinshed). And Henry VI, Henry V's one son, 'did abhorre of his awne nature, all the vices, as well of the body as of the soule, and from his verie infancie, he was of honest conuersacion and pure integritie, no knower of euill, and a keper of all goodnes' (Hall); 'by reason whereof, king Henrie the seauenth sued to Pope Iulio the second, to have him canonized a saint' (Holinshed). Tudor writers liked Tudor progenitors. But even Tudors kept too many people at court. On 24 May 1406, 40 servants were thrown out of Henry IV's house on the Commons' order; and on 8 December John Tiptoft, their speaker, was made royal household treasurer. The Parliament of 1449–50 offered £5,630 a year for seven years to meet Henry VI's costs; and it got rid of his minister Suffolk, whose head the friends of Jack Cade had. Henry V – the big winner at Agincourt – tended to get what he asked for. ('O God of warriors, England is indebted to Thee!')[87]

All but the last of the Lancastrian Henries had a son to succeed him. Henry VI lost the kingdom to, got the kingdom back from, then lost the kingdom again to Edward IV – heir to the house of York. But the usurpation did not last. When Edward's daughter Elizabeth married Henry Tudor, the Yorkist dynasty was put to rest. And Edward IV was, necessarily, remembered badly. There was a handful of bastards; there were several affairs. The Duke of Buckingham offered his successor a crown on the grounds that Edward's 'gredy appetite was insaciable, and euery where ouer al the realme intollerable.' He went on: 'for no woman was there any where yong or olde, riche or pore, whom he set his eie vpon, in whome he any thinge lyked either person or fauor, speche, pace, or countenance, but wtout any fere of god, or respect of his honour, murmure or grudge of ye worlde, he would importunely pursue hys appetite, and haue her.' The slander was international. To Dominic Mancini – the Italian poet – he was 'licentious in the extreme: moreover it was said that he had been most insolent to numerous women after he had seduced them, for, as soon as he grew weary of dalliance, he gave up the ladies much against their will to the other courtiers'; to Philippe de Comines – Louis XI's confidant – he was fonder than any another man of pleasure, '*speciallement aux dames et aux festes et banquetz et aux chasses*'. Edward seems to have won women in the usual ways. Elizabeth Woodville (says Mancini) became Edward's wife when the king 'placed a dagger at her throat, to make her submit to his passion', and she 'determined to so die rather than live unchastely with the king'; Jane Shore (says More) was tempted by 'hope of gay apparel, ease, plesure & other wanton welth'. He spent a lot to impress them. The *Black Book of the Household* put together in Edward's reign guessed a baron's income to be £500 and a knight's £100; Edward, on the other hand, spent £397 one month on wash basins, and £984 on a series of tapestries commissioned to show scenes of Nebuchadnezzar and Alexander. The Duke of Buckingham told Parliament the king took too many taxes and tallages, of which there was neuer end, & often time no nede'; but John Fortescue – who wrote political tracts for both Edward IV and Henry VI – thought the commons had no cause to complain. In England the king's agents might take goods for his house at 'a reasonable price;' in France they laid whole villages waste, 'paying not a penny for any of their own necessaries nor those of their concubines, whom they always carried with them in great numbers.' When Edward IV died in his prime, his death was variously attributed: to loss of the Dutch alliance to France; to eating too much; to his stepsons and other 'promoters and companions of his vices' who had 'ruined his health'; and to excessive interest in his barons' women. 'He would use himself more familiarly among private persons than the honour of his majesty required' – and might have been poisoned.[88]

Edward IV had the good luck to leave behind two legitimate sons. He had the bad luck to die in their minority. He made his younger brother Richard his sons' protector; and Richard – the Lancastrians said – had them suffocated in the Tower. There were lots of Lancastrian slanders. To Polydore Virgil – who wrote his *Anglica Historia* on Henry VII's commission – he had 'a sharp wit, provident and subtle, apt both to counterfeit and dissemble'; to Thomas More – who wrote his *History of King Richard III* to suit Henry VIII – he was 'close and secrete, a deepe dissimuler . . . not letting to kisse whome hee thoughte to kylle.' Some of them touched on sex. Holinshed said Richard III was 'greatlie giuen to fleshlie wantonnesse'; Hall said his 'sensuall appetite' took the place of 'right iustice'. More thought it enough that the king and his Yorkist brothers were all 'greate and statelye of stomacke, gredye and ambicious of authoritie, and impacient of parteners'; and even the sympathetic Croyland chronicler thought there was 'unseemly stress' on singing and dancing at Christmas, with 'many other matters which are not written down in this book, because it is shameful to speak of them.' Richard III conceived at least two children in sin. Katherine, a bastard daughter, was married off to the Earl of Huntingdon in 1484; John of Pomfret, a bastard son, may have been killed by Henry Tudor in 1499. But this Richard died heirless. He lost his wife's only son a year after winning the throne. A year later it did not matter. Henry Tudor beat him at Bosworth Field on 22 August 1485. Five months later, Henry married Edward's daughter. The wars of the roses were over. 'The red rose, the avenger of the white, shines upon us.' The king of England, for generations *primus inter pares*, was *rex imperator* again.[89]

Henry VII justified his acts by his descent from Lancastrian kings. But all of Edward III's legitimate lines had been wiped out by bastards' sons. Richard III had called Henry Tudor 'descended of bastard blood both of the father's side and of the mother's side.' On some grounds: Henry VII's father, Edmund Tudor, had been got on Katherine of Valois by Owen Tudor, after her husband Henry V died. But Katherine's will makes no mention of a second 'husband' Owen; and Henry VI, Katherine's one legitimate son, knew nothing of his mother's 'remarriage' until she died in his sixteenth year. Henry VII's other claim to Lancastrian ancestry was through his mother: Margaret Beaufort was Katherine Swynford's great-granddaughter. But Swynford was the lowly governess who bore John of Gaunt three bastard sons. They were not legitimized ('*excepta dignitate regali*') by Henry IV till 1407, and then only at the instigation of his upstart son. Legitimate or not, the new line was soon secured. Henry VII got nine heirs on Elizabeth of York.

Like other Tudor kings, this Henry was popular with Tudor chroniclers. Polydore Vergil could see that he had small eyes, few teeth, thin hair and a sallow complexion, but said he was 'remarkably attractive'; Erasmus, who dedicated a little poem on the royal brood to Henry's 8-year-old son, thought Henry VII 'indulgent' to others but 'strict' about his own behaviour: 'giving his citizens free rein, he keeps a tight rein on himself.' Francis Bacon – who wrote his *History of the Reign* for James I, who was Henry VII's great-grandson – said 'for his pleasures, there is no news of them'. He was sure Henry's laws were his 'preeminent virtue and merit'. But harsh: of 192 statutes in 7 parliaments, 35 were acts of attainder or restitution; in all, 138 were attainted for treason. As the Bishop of Rochester put it in his funeral oration, 'if any treason was conspired against him it came out wonderfully.'[90]

Nor was his son Henry VIII easily libelled. As the world knows, he had trouble getting an heir. Two divorces, six wives and a break with the pope were the result. But just one of his bastards got recognized. On Catherine of Aragon's lady-in-waiting, Elizabeth Blount, Henry got a son in 1519 ('you shall vnderstande, the kynge in his freshe youth, was in the chaynes of loue' – Hall). Bessie was given a dower and married off to lord Tailboys, made a member of the king's chamber; later she would marry Lord Clinton, made earl of Lincoln. Bessie's son (Henry's 'worldly juell' – Holinshed) was made Duke of Richmond and Somerset on 18 June 1525; he might have been considered for the succession – before Henry's divorce plan was put into effect. On his third wife, Jane Seymour, Henry finally got his heir in 1537. But tubercular Edward VI was dead at 16. The Tudor dynasty was carried on by two daughters – 'Bloody' Mary and her long-lived half-sister, Elizabeth I.

Henry VIII, like most kings, seems to have loved his wives. He called Catherine of Aragon, soon after he had married her, as 'my most dear and well-beloved consort, the princess my wife.' He closed a letter to Catherine Parr: 'No more to you at this time sweetheart . . . give in our name our hearty blessings to all our children. . . . Written with the hand of your loving husband, *Henry R.*' He started a letter to Jane Seymour, 'the bearer of these few lines from thy entirely devoted servant will deliver into thy fair hands a token of my true affection for thee.' And, in the 17 Vatican Library letters to Anne Boleyn, he was poetic. He sent her the spoils of his chase ('seeing my darling is absent I can not less do than to send her some flesh representing my name; which is hart flesh for Henry'); he compared her to summer ('the longer the days are the farther off is the sun and yet the hotter; so it is with our love, for although by absence we are parted it nevertheless keeps its fervency, at least in my case and hoping the like of yours'); he closed his letters '*im H Rex mutable*,' '*H ♥AB♥ R*,' and '*H aultre ♥AB♥ ne cherse R*' (H seeks ♥AB♥ no other R).

But, like most kings, Henry seems to have loved other women. Not everybody said so. He let Thomas Cromwell, his martyred chamberlain, know he had left Anne of Cleves 'as good a maid as he found her'; and Anne Boleyn had complained Henry '*nestoit habile en cas de soy copuler avec femme*' – was not good in bed. But he probably got there. There were stories about a sister of the Duke of Buckingham; there were stories about Mary Boleyn, a sister of Anne's (her 'young master Carey' might have been his son). There was a rumour about a Belgian lady Henry entertained in France; there was a rumour – told to the Abbot of Westminster – that he once met a man out on his horse, who had 'a pretty wench' riding behind him, and that Henry 'plucked down her muffler and kissed her, and liked her so well that he took her' and kept her. Everybody knew Henry was a patron of the arts. He had Hans Holbein run around the continent painting the faces of five prospective brides – but thought he had been misled about Anne of Cleves: 'say what they will, she is nothing fair'; he rebuilt Bridewell, Hampton Court and Whitehall, and built St James – though Henry's martyred chancellor, Thomas More, dared to say in his *Utopia* that it would be good if a king could 'live harmlessly on what is his own'. Nobody denied Henry liked to dance. Eustace Chapuys, the Habsburg ambassador, was a little bothered that Henry went out 'banqueting with ladies, sometimes remaining after midnight, and returning by the river' with musical instruments and chamber singers after Anne Boleyn's arrest; when Catherine Howard was arrested, he consoled himself with '26 ladies' at a feast. To Martin Luther, 'Juncker Heintze will be God and does whatever he lusts.'[91]

For all Henry's efforts, the Tudor line expired. None of his legitimate children left legitimate children. So in 1603 James I – Henry VIII's sister Margaret's son's daughter's son – descended from Scotland to take the throne. But in the half-century since its last king had died, England had changed. The crown was now perennially short on cash – from 1539 to 1553 alone, crown lands yielding £100,000 a year were sold. At the same time, merchants were getting rich. English ships, manned by the likes of Sir Walter Ralegh and Sir Francis Drake, had staked out New World territories since the 1570s; they came back loaded with money and set up new markets. In the Old World there were new markets as well – in the Levant, the Baltic, India and Russia. The Muscovy Company was established in 1553, the East India Company in 1600, the Massachusetts Bay Company in 1628. Customs revenues, still mostly from the cloth trade in Europe, rose from £29,500 in 1551 to £120,000 in 1606 on the great farm alone. Population had risen from under 3 million in 1550 to nearly 4 million at the end of the century. People were more literate – over half the men in London could read, and one-third of the men outside London could sign their names by

1640; they were better educated – there were more grammar schools, private schools and colleges. There were more metalworkers, more tanners and, above all, more textile makers. There were more men in the Commons, and they were more outspoken. Members started to insist on their right to debate all matters of state. Peter Wentworth was first sent to the Tower for defending free speech in 1576; he died there, for the same reason, in 1597.

The seeds of change sprouted under the Tudors; under the Stuarts they grew. Feudal revenues sank. Early in his reign James I bargained to end feudal tenures in return for an annual parliamentary subsidy, but the deal fell through. Then an Act of Parliament in 1656, under Lord Protector Oliver Cromwell, converted all lands held in tenure from the king into freehold lands. Income from trade, on the other hand, was up by leaps and bounds. In 1635, at £368,000, the customs yielded over half the exchequer's total receipts. Population rose again – from around 4 million at James I's accession to around 5 million at James II's deposition; literacy nearly doubled – from around 25 per cent in 1600 to near 50 per cent at the end of the century. As early as 1700, one-third of England's national product was accounted for by industrial and commercial jobs, another 27 per cent by government or domestic service and the professions, and just 40 per cent by farming. And, throughout the seventeenth century, Parliament continued to grow uppity. The Stuarts had serious differences of opinions with most of their parliaments. James I made his position clear at Whitehall in a 21 March 1610 speech:

> Kings are iustly called Gods, for that they exercise a manner or resemblance of Diuine power vpon earth: For if you will consider the Attributes to God, you shall see how they agree in the person of a King. God hath power to create, or destroy, make, or vnmake at his pleasure, to giue life, or send death, to iudge all, and to bee iudged nor accomptable to none: To raise low things, and to make high things low at his pleasure, and to God are both soule and body due. And the like power haue Kings

an opinion with which the Commons found exception. In, for example, their protestation of 18 December 1621, they wrote:

> that the ardous and urgent affairs concerning the King, state and defence of the realm, and of the Church of England, and the maintenance and making of laws, and redress of mischiefs and grievances which daily happen within this realm, are proper subjects and matteres of consel and debate in Parliament.

James disagreed, and told the Spanish ambassador: 'The House of Commons is a body without a head . . . ; I am surprised that my ancestors should ever have allowed such an institution to come into existence.' James I was disconcerted; his son, Charles I, was decapitated. Charles was done in on 30 January 1649. The tables had truly turned. By 3 September 1658 Cromwell himself was dead;

and on 8 May 1660 Charles II was recalled. But the Restoration did not last a generation. In 1688 James II, Charles II's younger brother and successor, was deposed in favour of his daughter Mary's husband William of Orange. William came to England at the invitation of the Lords and Commons. James I, the first of the Stuarts, had come to rule a kingdom; James II, the last Stuart, had lost it to parliament.[92]

It was around that time that the king's household, like other aristocratic households, shrank. Once, the houses of English kings held more men, and almost certainly more women, than any other houses in England. The earliest account of the royal household, the *Constitutio Domus Regis*, was written around 1136 about Henry I's reign. It lists major officers, and their recommended pay, in the chancery and chapel, steward's department, buttery, chamber, constabulatory and marshalsea. On the king's hunting staff alone, the *Constitutio* lists 4 hornblowers, 20 serjeants, a leader of the line-hound and a berner (hound-feeder), along with unnumbered knight-huntsmen, huntsmen, huntsmen of the hounds on the leash, brach-keepers (a 'brach' was a small scent-hunting hound), fewterers (greyhound keepers), wolf-hunters and archers. The staff of a medieval king was, in short, large. Half a century later Walter Map, one of Henry II's courtiers, asked how a king could 'keep in order thousands of thousands', when lesser men had so much trouble with smaller households. 'For in a hall that holds many thousand diverse minds there must be much error and much confusion; neither he nor any other man can remember the name of each individual, much less know their hearts.' They had a lot of help. A thirteenth-century *Book of Fees* lists dispensers, larderers, herb preparers, naperers, ushers, falconers, banner bearers and at least one (Henry de la Mare held this post) 'guard of the court strumpets'. In peacetime, the king's staff typically cost the government more than anything else. As Tout, in his *Administrative History*, maintained, 'the whole state and realm of England were the appurtenances of the king's household.' Then, slowly but surely, the court shrank.[93]

Under the Stuarts, it was still huge. James I was notoriously spendthrift: after 45 years of Elizabeth's celebrated parsimony, the number of ushers, grooms, carvers, cup-bearers, pages, messengers, gentlemen of the bedchamber and privy chamber went up. In the privy chamber, the number changed from 18 under Elizabeth to 48 under James; and another 200 'gentlemen extraordinary' were added to the court. Costs were commensurately high. The royal household cost £64,000 in 1603, £90,800 in 1607, £81,200 in 1612, £85,500 in 1617; £12,566 was assigned the royal cofferer alone; another £11,280 went to the prince's cofferer. The numbers are most explicit for Charles I: there were around 1840–60 servants in the second Stuart household. They included: 172 on the queen's staff, 202 on the staffs of Prince Charles and other royal children, 210 yeomen of the guard, 55 gentlemen pensioners, 60 on the great wardrobe staff, 580–620 in the king's chamber and its offshoots, 263 in the royal stables, 305 below stairs and another 195 servants' servants. Costs stayed high. On 18 December 1635, the king's total revenues stood at £618,379; £382,908 were assigned for the royal household, wardrobe, chamber, pensions and fees.

Things changed in the next century. George I's household held on the order of 950. There were 640–66 in the chamber and its offshoots (not counting an unspecified number of 'gentlemen and boys of the choir' in the royal chapel); there were 31–7 in the bedchamber; there were 157 below stairs (from the Lord Soteward to wine porters, turnbroaches, to the ice house yeoman); there were another 104–9 in the stables (the number of grooms ranging from 13 to 18). But by 1806, when the future George IV was setting himself up at Carlton House, a staff of 40 indoor servants seemed large. Various gentlemen of the household, a treasurer, a keeper of the privy purse, a vice treasurer, 2 clerks, 2 surgeons, 4 pages of the backstairs, 3 pages of the presence, 2 grooms of the chamber and 7 musicians boarded out. Indoor servants included a housekeeper, a wardrobe keeper, a maître d'hôtel, an inspector of deliveries, 9 housemaids, 4 cooks, 3 watchmen, 2 kitchen boys, a kitchen maid, 2 confectioners, 2 cellarmen, 2 coal porters, a coffee-room woman, a silvery scullery woman and a table decker. The bill for monthly groceries was just under £500; the wine bill was £1,118. There was enough to do a few parties. Shelley, among others, thought them regressive. At one 19 June 1811 fête, according to the 21 June *Morning Chronicle*, George wore scarlet, gold lace and the star of the Order of the Garter; he sat at the head of a 200-foot table; and from him, 6 inches above the table, flowed a canal from a fancy silver fountain, 'its faintly weaving, artificial banks' covered with 'green moss and aquatic flowers', the water filled with gold and silver fish. As the poet put it to his friend Elizabeth Hitchener, in disgust,

> It is said that this entertainment will cost £120,000. Nor will it be the last bauble which the nation must buy to amuse this overgrown bantling of Regency. How admirably this growing spirit of ludicrous magnificence tallies with the disgusting splendours of the stage of the Roman Empire which preceded its destruction! Yet here are a people advanced in intellectual improvement wilfully rushing to a revolution, the natural death of all great commercial empires, which must plunge them in the barbarism from which they are slowly arising.

Another hundred years later, after Victoria had come and gone, her son Edward VII still kept a staff of over a hundred at Marlborough House. He took 33, as prince, on an 1868 trip to Paris; he took 31 on the royal yacht when he visited his sister in 1901. Still, 'Bertie' was living on an income substantially reduced from George IV's. Edward, as Prince of Wales, lived on about £65,000 a year; Prince George had got around £125,000. Since Edward, the numbers have continued to go down. Nevertheless, kings have kept having sex. Though over time – and this is the point – opportunities seem to have been drying up.[94]

In 1598 James IV of Scotland offered his young son Prince Henry a few words of advice. The *Basilicon Doron* was published in Edinburgh and London in 1603, around the time its author took the English throne as James I. Among other things, it advised against fornication – 'Yee must keepe your bodie cleane and vnpolluted, till yee giue it to your wife, whom-to onely it belongeth ... *The fornicator shall not inherite the Kingdome of heauen.*' It also advised against adultery – 'I trust I need not insist here to disswade you from the filthy vice of adulterie.' And it was disapproving of homosexuality. Of 'some horrible crimes that yee are bound in conscience neuer to forgiue', 'sodomie' was one.

If other writers are right, James I was a hypocrite on all counts. To Sir John Oglander, who kept *A Royalist's Notebook*, James 'loved young men, his favourites, better than women'. James loved in particular James Hay – who was 'made for a courtier' and married off to the Earl of Northumberland's heiress and daughter. He loved Robert Carr – the athletic Scot who was made Viscount Rochester in 1611 and Earl of Somerset later. And he loved most of all George Villiers, the younger son of a Lecistershire knight made a viscount in 1616, an earl in 1617, a marquis in 1618, and Duke of Buckingham in 1623. Again, men may have had sex with men; they almost certainly had it with women. To Sir Edward Peyton – an MP under the first two Stuarts, who lost a position in Cambridge thanks to the Duke of Buckingham – George Villiers 'by his greatness vitiated many gentile and noble virgins in birth, though vitious for yeelding to his lust; whose greatness opened the door to allure them more.' To Anthony Weldon – who was one of James's clerks of the Board of Green Cloth, got dismissed for a satire, then sided with Cromwell in the civil wars – the king was fat, weak-legged and never washed his hands; he had big eyes 'ever rowling after any stranger came in his presence'; and he tended to walk in circles, 'his fingers ever in that walk fidling about his cod-piece.'

To Simonds D'Ewes – who was made a baronet by Charles I in 1641, and lost his seat in the House of Commons in 1648 for having too much regard for the king – the trouble was the nobility. They were all 'prostituting their bodies to the intent to satisfy and consume their substance in lascivious appetites of all sorts.' Still, D'Ewes was glad in his *Autobiography* to see that when James I rode from Whitehall to Westminster in 1621, 'the windows were filled with many great ladies.'

Lucy Hutchinson – the wife of a colonel in Cromwell's New Model Army – thought the problem was the monarchy. 'The court of this king was a nursery of lust and intemperance'; there were 'fooles and bawds, mimicks and catamites'; James's Scots 'were faine to invent projects to pill the people, and pick their purses for the maintenance of vice and lewdnesse.' Francis Osborne – the Earl of Pembroke's master of horse who, along with his benefactor, opposed the king in the wars – agreed. 'This nation was rooted up by those Caledonian bores.' That they preyed on English ladies 'none can doubt that hath but the opportunity to peepe into a court, where the love of women is found a consequence of the favour of the prince.' An epigram 'everywhere posted' read:

> They beg our lands, our goods, our lives,
> They switch our nobles, and lye with their wives.

And the Stuarts were skewered on stage. John Fletcher, whose *Humorous Lieutenant* first played in 1619, had Leucippe, his 'Agent for the King's Lust,' complain she needed 'some twenty young and handsome,/As also able maids, for the Court service.'[95]

Charles I, James I's son, was short; he stammered; and he was devoted to his family – preserved for posterity in beautiful portraits by Van Dyck. Charles had failed (spectacularly) to make an alliance with one absolutist's daughter – the Infanta Maria, child of Philip III of Spain. But he had managed (respectably) to marry another – Henrietta Maria, daughter of Henry IV of France. After which, he was kind enough to keep mistresses, and issue, out of sight. This Charles is supposed to have dressed modestly, eaten moderately, and had very little to drink. To sympathetic Edward Hyde – who wrote his great *History of the Rebellion* to please Charles I, became Charles II's Lord High Chancellor, and was father-in-law to James II – Charles I was 'saint-like' and a 'blessed martyr'; he offered 'so great an example of conjugal affection, that they who did not imitate him in that particular did not brag of their liberty'; and he died 'the worthiest gentleman, the best master, the best friend, the best husband, the best father, the best Christian, that the age in which he lived had produced'. Even to unsympathetic Lucy Hutchinson, Charles was 'temperate, chast and serious' and 'most uxorious' – though 'the nobility and courtiers, who did not quite abandon their debosheries, had yet that reverence to the King to retire into corners to practise them'. Charles's courtier Robert Read, in a 23 January 1640 letter to his cousin Thomas Windebank, agreed. 'We keep all our virginities at court still; at least we lose them not avowedly.'[96]

The first two Stuarts, James I and Charles I, were well provided with legitimate children. The last two Stuarts, Charles II and James II, were not. James II got an heir on the eve of his deposition; and Charles II got no heirs at all. Neither James I nor Charles I had any recognised bastards; between them, Charles II and James II had 20 or more.

Sex was not kept under wraps at the Restoration court. There were several satires. An anonymous 'Essay of Scandal' of the summer of 1681 blamed Charles II's poverty on his prodigality with females. 'Why art thou poor, O King? Embezzling cunt,/that wide-mouthed, greedy monster, that has done't.' There were poems. Andrew Marvell's 'The Kings Vowes' read, in part,

> But what ever it cost me, I will have a fine Whore,
> As bold as Alce Pierce and as faire as Jane Shore;
> And when I am weary of her, I'le have more.

> Which if any bold Commoner dare to opose,
> I'll order my Bravo's to cutt off his Nose,
> Tho' for't I a branch of Prerogative lose.

> Of my Pimp I will make my Minister Premier,
> My Bawd shall Embassadors send farre and neare,
> And my Wench shall dispose of the *Congé d'eslire*.

There were histories. To Gilbert Burnet – the Bishop of Salisbury – Charles confided, 'he was no atheist, but he could not think God would make a man miserable for taking a little pleasure out of the way.' And there were memoirs. Anthony Hamilton's *Memoirs of the Court of Charles the Second by Count Gramont* said the king 'was often the dupe, but oftener the slave, of his engagements.' He liked the greater and 'all the lesser mistresses he had in various parts of the town.' After all, 'he was the wittiest man in the world and he was King. These are no mean qualifications.'

Charles II was fond of his wife. And she was fond of him. Katherine of Braganza seems to have done what she could to provide an heir ('as all the happiness of her life depended upon that blessing, and as she flattered herself that the king would prove kinder to her if heaven would vouchsafe to grant her desires'); but she had no luck. So the king devoted himself to several broods of bastards. The best known were born to Barbara Villiers, whose husband Roger Palmer was made Earl of Castlemaine and who was later made Duchess of Cleveland in her own right; the actress Nell Gwyn; Elizabeth Killigrew, whose daughter Charlotte Jemima Henrietta Maria was married first to the second Earl of Suffolk then to the second Earl of Yarmouth; Catherine Pegge, whose son 'Don Carlo' was made Earl of Plymouth, and whose daughter Catherine died young or became a Dunkirk nun; Louise de Querouaille whose son, Charles Lennox, was made Duke of Richmond; and Lucy Walter whose son, later made Duke of Monmouth, was born at Rotterdam in 1649. (Prince William of Wales, the future king of Britain, is descended from Charles II through his mother, Diana, Princess of Wales, who was a descendant of Charles II's son, the Duke of Grafton, by Barbara Villiers.) Many more of the king's women received royal pensions. The Commons accordingly complained. Of all Charles's children, just one made a serious bid for succession. Monmouth seems to have taken after his father. 'He was particularly beloved by the king, but the universal terror of husbands and lovers.'

Again, the Commons was complaining with cause. One day in October 1683, Pepys's friend John Evelyn paid the Duchess of Portsmouth a visit. When he left, 'surfeiting of this', he went 'contentedly home to my poore, but quiet Villa.' The duchess's dressing room was rich. It had already been rebuilt 'twice or thrice' to suit her tastes; there was 'French Tapissry', there were Japanese cabinets and 'Skreenes', there were vases, furniture and plate 'all of massive silver, & without number'; there were his majesty's best paintings – of palaces, stags and exotic birds. 'Lord, what contentment can there be in the riches & splendour of this world, purchas'd with vice & dishonour.' Charles's favourites competed for his favours. One day the Queen, Lady Castlemaine and Frances Stuart all wanted a ride through Hyde Park in the king's new carriage: Count Gramont had special-ordered it from Paris at a cost of 1,500 louis, and it was 'the most elegant and magnificent calash that had ever been seen.' Lady Castlemaine, who was pregnant, threatened to miscarry if she was not the king's first passenger; Frances Stuart 'threatened that she never would be with child' if she was not picked. Anthony Hamilton says 'this menace prevailed'. (Lady Castlemaine was incensed, though the trip 'cost her rival some of her innocence'.) To Evelyn, who made an entry in his diary the day Charles died, he was 'an excellent prince doubtlesse had he ben lesse addicted to Women, which made him uneasy & allways in Want to supply their unmeasrable profusion' – to the detriment, he added, 'of many indigent persons'. And to Bishop Burnet, 'the ruine of his reign, and of all his affairs, was occasioned chiefly by his delivering himself up at the first coming over to a mad range of pleasure.'[97]

Charles II was succeeded by the Duke of York, his younger brother James II. James too was fond of his first wife Anne Hyde, but none of her four sons lived. Anne died in 1671; till then, Anthony Hamilton says, James had fun. And the duchess suffered. 'In the face of his flirtations it was her duty to be patient until it pleased Heaven for him to repent.' He ran around and lost weight; she took after Jack Sprat's wife.

> The Duchess of York was one of the highest feeders in England: as this was an unforbidden pleasure, she indulged herself in it, as an indemnification for other self-denials. It was really an edifying sight to see her at table. The duke, on the contrary, being incessantly in the hurry of new fancies, exhausted himself by his inconstancy, and was gradually wasting away.

The duchess was reluctant to take a lover herself, even if the duke, 'having received the favours, or suffered the repulses of all the coquettes in England', was busy seducing her Maids of Honour – hand-picked by Anne Hyde herself to include 'only the most beautiful' young women in her kingdom. 'A woman must have superhuman patience, or be incredibly resigned.' On his accession, James seems to have sobered up somewhat. According to Burnet, 'the King did, some days after his coming to the Crown, promise the Queen and his Priests, that he would see Mrs *Sidley* no more, by whom he had some five children. And he spoke openly against lewdness, and expressed a detestation of drunkenness.' As often happened, his acts were not quite as good as his words. Mrs Sidley was ordered to leave her lodgings at Whitehall; 'yet the King still continued a secret commerce with her.'[98]

When James II married Mary of Modena – the 15-year-old daughter of a Catholic duchess given a £90,000 dowry by Louis XIV – Parliament protested and papal effigies were burned. And when, after 15 years and several dead children, she gave James an heir, it was more than the opposition could bear. James, Mary and their boy were packed off to France and never came back. And William of Orange arrived from Holland to rule Britain with his wife – another Mary, James II's first daughter by Anne Hyde. Thirteen years later they were succeeded by James II's second daughter, named after her mother, Queen Anne. But it was the bad luck of these queens to die without any surviving children; so the Stuart line expired in 1714. Then the Georges came. James I's great-grandsons were imported from Hanover. Parliament had invited them in, and the monarchy would never be the same.

George I was not altogether clean. He liked to be known as a ladies' man. According to Horace Walpole – his minister Robert Walpole's son – the first Hanoverian's women included one he made Countess of Darlington, 'by whom he was father of Charlotte Viscountess Howe, though she was not publicly avowed.' Towards the end of his life, George gave an apartment in St James to another, Miss Anne Brett, who 'was to have been created a countess', but afterwards married Sir William Leman. He told a third, made Duchess of Kendal, that he would (if he could) come back to her after the end of his life. He died; a bird flew in through the duchess's window, and 'she believed it was the King's soul, and took the utmost care of it.' This mistress, also known as Melusine von der Schulenberg, was George I's great love. By her he had had three natural daughters: Anna Louise, married to Ernst August Philipp von dem Bussche-Ippenburg, divorced, and made Reichsgräfin von Delitz in her own right; Petronella Melusine, made Countess of Walsingham in 1722, and married to Philip Dormer Stanhope, Lord Chesterfield, in 1733; and Margarethe ('Trudchen') Gertrud, George's favourite, who died young. None was ever formally legitimised or publicly acknowledged; all were officially registered as children of Melusine's sisters; and each was known to England as a niece of Melusine.[99]

George I had just one son and one daughter by Sophia Dorothea of Brunswick and Zelle – the wife he divorced in 1694. But he had terrifically prolific scions. His son George II had, by Caroline of Brandenburg-Anspach, three sons and five daughters; and his great-grandson George III, by Sophia Charlotte of Mecklenburg-Strelitz, had six daughters and nine sons.

George II liked women, like his father. We know that partly from Lord Hervey – his queen's confidant. George seems to have told his wife all about his affairs. In his *Memoirs of the Reign of King George II*, Hervey says the king was so honest when he courted Madame Walmolden that he 'acquainted the queen by letter of every step he took in it, of the growth of his passion, the progress of his applications, and their success', down to the details of what he paid – a thousand ducats at first, which Hervey regarded as a trivial price. Horace Walpole confirms. He says in a footnote that 'the King was so communicative to his wife, that one day Mrs Selwyn, another of the Bedchamber women, told him he should be the last man with whom she would have an intrigue, because he always told the Queen.'

Caroline, in any case, put up with her husband's women – like 46-year-old Mrs Howard, wife of the Earl of Suffolk's younger brother, whose husband the king eventually pensioned off at £1,200 a year. In this case the queen, says Hervey, 'knowing the vanity of her husband's temper, and that he must have some woman for the world to believe he lay with, wisely suffered.' Sometimes, George even asked her for help. In one of the 40-plus page letters he sent her on a 1735 trip to Hanover, George told her 'to contrive, if she could, that the Prince of Modena, who was to come the latter end of the year to England, might bring his wife.' In the original French: 'Je suis sûr, ma chère Caroline, vous serez bien aise de me procurer, quand je vous dis combien je le souhaite.' Robert Walpole, George's minister and Horace's father, advised the queen to put Lady Tankerville in the king's quadrille; after all, 'if the king would have somebody else, it would be better to have that somebody chosen by her than by him'; and Lady Tankerville 'was a very safe fool'. Even Blackburne, the Archbishop of York, was pleased 'that Her Majesty was so sensible a woman as to like her husband should divert himself.' And Caroline's daughter, the Princess Royal, approved. To Lord Hervey she said, 'I wish, with all my heart, he would take somebody else, that Mamma might be a little relieved from the ennui of seeing him for ever in her room.'

Still, this George did not get everything he wanted. The Countess Mary Cowper, who kept a diary, was sure she was admired but chose to hold on to her honour. When Madame Walmolden, on the other hand, gave in, she got £50,000 and another £50,000 for her husband – at least, the Earl of Egmont thought. Pregnancy probably raised the price. In 1735, Lord Hervey was sure, Madame Walmolden's announcement 'had extremely increased his fondness'; the birth of a son the next year 'had very much whetted his impatience to return.' Two years later, when Lady Deloraine had taken Madame Walmolden's place in the king's life, Robert Walpole asked her who had fathered her 1-year-old son. 'To which her Ladyship, before half-a-dozen people, without taking the Queen at all ill, replied, "Mr. Wyndham, [her husband,] upon honour"; and then added, laughing, "but I will not promise whose the next shall be."' She confided in private that 'she thought old men and kings ought always to be made to pay well'; besides, 'she was sure her husband . . . would not take it at all ill.' Fruit was not, though, the only object of George II's passions – not, at least, where Mrs Howard or Madame d'Elitz were concerned. Madame d'Elitz, Lady Chesterfield's sister, had had a thousand lovers, and 'was said to have been mistress to three generations of the Hanover family.'[100]

George III is remembered as a family man. He thought himself one. He was fond of his wife, and said of Charlotte (in a 14 August 1780 letter) to his son: 'I can with truth say that in nineteen years I have never had the smallest reason but to thank Heaven for having directed my choice among the Princesses then fit for me to marry, to her; indeed I could not bear up did I not find in her a feeling friend to whom I can unbosom my griefs.' And George was fond of his children. As Mrs Delany said (in a 10 October 1779 letter) to her niece, Miss Port: 'I never saw more lovely children; nor a more pleasing sight than the King's fondness for them, and the Queen's; for they seem to have but *one mind*, and that is to make everything easy and happy about them.' George III's fidelity seemed obvious even to Byron – whose *Vision of Judgement* allowed his 'household abstinence; I grant/His neutral virtues, which most monarchs want;/I know he was a constant consort; own/He was a decent sire, and middling lord.' The king and queen and their fifteen children – only two of whom died in infancy, eight of whom lived past 70 – seemed the picture of domestic tranquillity; and so, thanks to Thomas Gainsborough, they remain.

George III seems to have wanted women only when he was crazy. He lost the colonies, his mind, and the monarchy – and, in his son's regency, called out for a few ladies. He babbled about Lady Pembroke; he threatened 'to keep a mistress and several times has declared that since he finds Lady Yarmouth will not yield to his solicitations he will make love elsewhere'; he made a speech to several women on his yacht, 'with peculiar emphasis and strength of voice', and said in particular to Mrs Drax: '"You look very well, very well indeed, dear lovely Mrs Drax, how I should like to [ ] you."'[101]

George IV was unlike his father. He was unhappily married – to Caroline of Brunswick-Wölfenbuttel – and he had just one legitimate daughter, Charlotte, who predeceased her father in 1817. George IV, like George I, had few legitimate children; both took credit for several bastards.

Girls had always been attracted to George IV. At 17 he was said to have seduced one of his mother's maids of honour, written love letters every day to the Duke of Hamilton's great-granddaughter, and fallen violently in love with – and promised £20,000 on coming of age to – a pretty actress, Mary Robinson. She got £5,000 and a £500 annuity in the end. Other affairs followed: names include Lady Augusta Campbell – the Duke of Argyll's daughter; Lady Melbourne – who may have given birth to his son; Lucy Howard – whose son George, supposed to have been his, died young; Louise Hillisberg – the dancer; Harriet Wilson – the courtier; Elizabeth Billington – the singer; Mrs Anna Maria Crouch – another singer; Mme de Meyer; the Countess of Salisbury; the Countess of Jersey; Count Karl August von Hardenburg's wife; the Earl of Massereene's wife; Sir Charles Bamfylde's wife; Lord Hertford's wife; Lord Conyngham's wife. George may or may not have succeeded with the Duke of Devonshire's wife. Most husbands were willing and well paid. On 13 January 1823 Princess Lievenen wrote to Prince Metternich that the king had complained about Lady Conyngham: 'You see how she takes advantage of her position to push her family. Oh, she knows very well she is well off.' But George was not always irresistible. Lady Bessborough described his efforts to seduce her in a letter to her lover, Lord Gower:

> He threw himself on his knees, and clasping me round, kissed my neck before I was aware of what he was doing. . . . Then mixing abuse of you, vows of eternal love, entreaties and promises of what he would do – he would break with Mrs. F. and Lady H., I should *make my own terms!!* I should guide his politics, Mrs. Canning should be Prime Minister. . . . Then over and over and over again the same round of complaint, despair, entreaties and promises, and always Mr. Canning.

George loved many women, but Maria Fitzherbert was the love of his life. She was the widow of Thomas Fitzherbert and Edward Weld; she was the granddaughter of a northern Catholic, Sir John Smythe; and, on 15 December 1785, she probably (secretly) became George IV's wife.

Most of all, though, George IV seems to have been fond of himself. According to the 4–7 June 1791 edition of *St. James' Chronicle*, he showed up for his father's birthday in bottle-green and claret-red striped silk coat and pants, a silver tissue waistcoat and a lot of diamonds – on his waistcoat, his pants, his epaulette and his sword. Those tastes stuck. Thirty years later, Benjamin Haydon watched his coronation:

> Something rustles and a being buried in satin, feathers and diamonds rolls gracefully into his seat. The room rises with a sort of feathered, silken thunder! Plumes wave, eyes sparkle, glasses are out, mouths smile, and one man becomes the prime object of attention to thousands! The way in which the King bowed was really monarchic! As he looked towards the Peeresses & Foreign Ambassadors, he showed like some gorgeous bird of the East.[102]

When George IV died without heirs, his brother William succeeded. William IV was, for many years, unprepared. George had planned to leave the kingdom to his brother Frederick, Duke of York; but Frederick died in 1827, and three years later William was made king. He had spent a life at sea. Packed off to Hanover at 18, he had complained of the locals in a letter to his big brother George: 'Oh! I wish I was returned. . . . England, England for ever and the pretty girls of Westminster; at least, to such as would not clap me or pox me every time I f***ed.' William fell in love with his cousin, Princess Charlotte of Mecklenburg-Strelitz; he may or may not have got a bastard on Caroline von Linsigen, a Hanoverian general's daughter.

He found other women in ports. To Lord Elphinstone on 3 January 1787 William wrote: 'excuse me for not having wrote this long time, as I have been in constant round of dissipation from my first arrival in the West Indies and I am afraid it will continue as long as I remain in these seas;' to his brother George he wrote on 20 May of the same year that he had had mercury for 'a sore I had contracted in a most extraordinary manner in my pursuit of the *Dames des Couleurs*.' William had dinner in Halifax with John Wentworth, the governor, and is supposed to have enjoyed his wife; otherwise, his friend William Dyott put in his diary, 'he would go into any house where he saw a pretty girl.' He brought one pretty girl from Mayfair to Richmond; he made another's father – a Plymouth merchant – Agent Victualler at £200 a year. William kept writing home to George, complaining on 26 October 1788, 'I am sorry to say I have been living a terrible debauched life, of which I am heartily ashamed and tired.'

William got ten bastards on 'Mrs.' Jordan. She was the bastard daughter of Irish parents, the discarded mistress of a lawyer and a theatre manager, and maybe London's most famous actress. In the spring or summer of 1791, 29-year-old Dorothy Jordan and her four illegitimate children moved in with William; by 1807 she had borne him his ten. George IV was made regent in 1811, when Dorothy was nearly 50; William settled £4,400 a year on her and looked about for a wife. He married Princess Amelia Adelaide Louisa Theresa Caroline of Saxe-Meiningen in 1818. They had no children. So when William died in 1837, the monarchy passed to his dead younger brother Edward's daughter, Victoria. Victoria ruled Britain for 64 years.[103]

When she died, she left the kingdom to Albert Edward, her oldest son. It was 1901, and England had continued to change. Prince Edward at 19 had started feeling his oats with Nellie Clifden, another actress, back in 1861. The details were familiar; the disapproval was new. Victoria's husband, the Prince Consort Albert, was upset. He wrote to Edward that Nellie 'can drag you into a Court of Law to force you to own it and there with you (the Prince of Wales) in the witness box, she will be able to give before a greedy Multitude disgusting details of your profligacy for the sake of convincing the Jury.' He might be cross-examined by an 'indecent attorney'; he might be hooted down 'by a Lawless Mob!! Oh, the horrible prospect.' Horrible enough that by December the Prince Consort was dead of typhoid fever, and Victoria was sure 'that Bertie was to blame'. As she wrote to the Crown Princess of Prussia, her firstborn daughter, two weeks later:

'I feel daily, hourly, something which is too dreadful to describe. I never can or shall look at him without a shudder.' In another letter she went on: 'Pity him, I do . . . But more you cannot ask. This dreadful, dreadful cross kills me!'

Edward's attitude was different from his mother's. A few years later, he defended himself: 'In every country a great proportion of the aristocracy will be idle and fond of amusement . . . We have always been an Aristocratic Country, and I hope we shall always remain so, as they are the mainstay of this Country, unless we become so Americanized that they are swept away.' Bertie practised what he preached. Over the years, there were rumours of his affairs in St Petersburg, Moscow, Paris, New Delhi; in Paris alone, French detectives traced him to meetings with the Comtesse Edmond de Pourteles, the Baronne Alphonse de Rothschild, the Princesse de Sagan, a Mme Kauchine, the widow Signoret, a Dame Verneuil, Baronne de Pilar, and unidentified ladies at the Hôtel Scribe and his favourite brothel, Le Chabanais. In England, there were chorus girls in sedan chairs at his friend Wynn-Carrington's; there were what Francis Knollys called his 'actress friends', what Lady Geraldine Somerset called 'his troop of fine ladies', what Edward Hamilton called 'H.R.H.'s virgin band'. For a while, according to the Duke of Cambridge, Edward took 'a strange new line', taking 'young girls and discarding the married women'; but in the long run he liked both. There were (probable) affairs with the wife of the Prefect of Moscow, the wife of Lord Wavertree, the wife of Lord Brooke; there were (notable) affairs with Lillie Langtree (the actress), Daisy Warwick (the countess) and Alice (Mrs George) Keppel. 'A cloud of blue bottle flies constantly buzzed round the king.' To Henry James, who was sorry to see 'little mysterious Victoria' go, Edward was an 'arch vulgarian', 'Edward the Caresser'. In spite of which, his wife was sure, 'he always loved me the best.'[104]

By his wife, Alexandra of Denmark, Edward VII had six children. His son George V succeeded him in 1910. He too was happily married – to a distant cousin, Mary of Teck. On 16 July 1893, a few months after his wedding, George wrote his friend, Flag-Lieutenant Bryan Godfrey-Fassett, 'All I can say is that I am intensely happy, far happier than I ever thought I could be with anybody. I can't say more than that.' To his wife, in August 1894, George wrote, 'I know I am, at least I am vain enough to think that I am capable of loving anybody (who returns my love) with all my heart and soul, and I am sure that I have found that person in my sweet little May.' In the next twelve years, she bore him five sons and a daughter. And they all lived – said John Gore in the 'personal memoir' he wrote 'at the request of Their Majesties, the King and Queen Mary' – in cozy contentment at York Cottage.

At least until 1910. Two rumours flared up on the king's accession. One, on George's intemperance, faded away. The other – that the king as a young man in the Mediterranean Fleet had married Admiral Sir Michael Culme-Seymour's daughter in Malta, that children had been born, and that wife and children had been abandoned when George found himself in line for the succession – was stopped by a signed denial in 1911.[105]

Britain continued to change, and so did Balmoral. Edward VII's friend, Lord Esher, noted the difference in his diary at the beginning of George's reign: 'It is altogether different here from former years. There is no longer the old atmosphere about the house – that curious electric element which pervaded the surroundings of King Edward. Yet everything is very charming and wholesome and sweet.' The six kids ran about the house; the queen liked to knit at night. But the king was content. To Count Albert Mensdorff, another Edward VII crony, George confided: 'We have seen enough of the intrigue and meddling of certain ladies. . . . I'm not interested in any wife except my own.' Lloyd George, future Prime Minister, wrote to his wife Margaret on 8 September 1910 after his first trip to Balmoral, 'The King is a very jolly chap but thank God there's not much in his head. They're simple, very, very ordinary people, and perhaps on the whole that's how it should be.' A year later, after another visit, Lloyd George was more careful. 'I am not cut out for court life. . . . Everybody very civil to me as they would be to a dangerous wild animal whom they fear and perhaps just a little admire.' By the eve of the First World War, the king was resentful. The Cabinet was neglecting to inform him of its proceedings. George's private secretary Lord Stamfordham complained, 'His Majesty is deeply pained at what he regards as not only a want of respect, but as ignoring his very existence.' Then in January 1924, a few years after the war was over, Labour and Liberals joined to defeat the Conservatives by 72 votes; and James Ramsay MacDonald replaced Stanley Baldwin as Prime Minister. As J. R. Clynes, the new Lord Privy Seal, declaimed: 'Fortune's Wheel had brought MacDonald the starveling clerk, Thomas the engine driver, Henderson the foundry labourer and Clynes the mill-hand, to this pinnacle beside the man whose forebears had been kings for so many splendid generations. We were making history.'[106]

Last and least, there was George VI. He came to the throne following his father's death and brother Edward VIII's abdication in 1936; when he died in 1952, he left England to the first of his two daughters, Elizabeth II. Like his remote ancestor, Charles I, George VI stammered and was shy; like his immediate ancestor, his father, he liked a quiet life. To one of his brother's faster friends, George VI and his wife Elizabeth were rather 'prim and proper'; they liked a low social profile.

Nor were they particularly powerful. When Europe was on the eve of its Second World War, the King of Britain felt he ought to get involved. He wanted, in particular, to talk to Hitler. The Prime Minister disagreed. In a secret report sent to Hitler in 1936, the Duke of Coburg wrote, 'To my question whether a discussion between Baldwin and Adolf Hitler would be desirable for future German–British relations, he replied in the following words: "Who is King here? Baldwin or I? I myself wish to talk to Hitler." ' Very few shared his point of view. The king, being Hanoverian, might have been pro-German. He was naive. And he was incidental.[107]

Conclusion

What put an end to polygyny in Britain?

Not the Christian Church. Old Testament men are consistently polygynous – from patriarchs like Abraham (who had three women) to kings like Solomon (who had a thousand). In the New Testament, Jesus like Paul was an advocate of chastity; but neither had an obvious immediate effect. Constantine's Edict of Milan essentially established Christianity as Rome's state religion in 313; the church chastised fornicating clerks from as early as the Council of Nicaea of 325; and canon law discouraged concubinage for hundreds of years – all, again, apparently to little effect. Puritans in England admonished promiscuous men; they were succeeded by the Restoration court. Rich men in Rome, throughout the Middle Ages, and in modern England continued to do as rich men had always done. They *married* monogamously: that is, they took one rich, heir-bearing wife at once. But they *mated* polygynously: that is, they had sex with as many women as they could afford.[108]

Nor companionate marriage. In early-modern England, husbands were often told to love their wives. John Stuart Mill, in his nineteenth-century essay on *The Subjection of Women*, put it well: marriage approaches ideal 'when each of two persons, instead of being a nothing, is a something.' Mary Shelley's mother, a century earlier, had complained that 'fondness is a poor substitute for friendship!' A wife wanted, and deserved, respect. But wives had always had that. When, for instance, Cicero wrote to Terentia, his noble first-century BC Roman wife, his letters were full of affectionate regard: 'Light of my life, for whom I yearn'; 'I seem to see your very face, and so I break down and weep'; 'I assure you, my dearest, that my desire is to see you as soon as possible and die in your arms.' Wives – mothers of heirs – had always been well honoured, and often well loved, by their husbands. The matter of marriage was never sexual fidelity. The matter of marriage was property.[109]

What put an end to British polygyny? *Democracy*. Power is – as it always was – a means to sex. A man with less power is a man with fewer favours to offer; and a man with fewer favours to offer is less attractive to women, and to their husbands. Powerful men in England filled their households with young, unmarried women, had sex with them – and then, occasionally, got them good husbands. Powerful Englishmen had sex, too, with their subordinates' wives – and then, occasionally, got their husbands promotions. When men in England, then on the continent, then all over the world lost power, their maids moved out, and their mistresses' husbands were less easily paid off.

Public life *is* private life. The point of power is sex. That is true not just of British history, it is true of human history. And it is true of natural history. From paper wasps (and other insects), to groove-billed anis (and other birds), to chimpanzees (and other mammals), to outcompete everybody else is to outreproduce everybody else. People are no exception. From hunters, to horticulturalists and herders, to farmers, men who have out-fought other men have won better habitats, more food and more sex. And they have almost certainly produced more children as a result.[110]

For a few million years, we hunted or scavenged for a living, and lived in fairly free democracies. Then, for the last few thousand years, many of us farmed for a living, and lived in despotisms. The tide turned, in England, in the last few hundred years. Despotism gave way, again, to democracy. And polygyny started to give way to monogamy.

Acknowledgements

It is a privilege to thank Christopher Brooke, Alan Macfarlane, William McNeill and Richard Wall for taking the trouble to read and make remarks on this chapter. I am especially grateful to Lawrence Stone, for whom my admiration must be obvious, for taking the time over the last few years to correspond. This chapter is dedicated to Sarah Blaffer Hrdy. A friend in need is a friend.

Notes

1 T. H. Hollingsworth, 'Demography of the British peerage', supplement to *Population Studies*, xviii (1957), 48.
2 Ibid., 32, 48–9.
3 Livy, *History of Rome*, i, 57–60 (translated by Aubrey de Sélincourt, Harmondsworth, 1971); Suetonius, *Augustus*, 71 (translated by Robert Graves, revised by Michael Grant, Harmondsworth, 1982); Tacitus, *Annals*, vol. 10 (translated by Michael Grant, Harmondsworth, 1989); Dio Cassius, *History*, lix, 4.1 (translated by Earnest Cary, Putnam, 1917); L. Friedlander, *Roman Life and Manners under the Early Empire* (Leonard Magnus's translation, London, 1908), vol. 1, 43.
4 Overall, see L. Betzig, 'Roman polygyny', *Ethology and Sociobiology*, xiii (1992), 309–49.
5 *Digest*, 27.1.2.2–3, 50.5.2.5; also B. Rawson, '*Spurii* and the Roman view of illegitimacy', *Antichthon*, xxiii (1989), 10–41 and L. Betzig, 'Roman monogamy', *Ethology and Sociobiology*, xiii (1992), 351–83.
6 G. Duby, *Medieval Marriage* (Baltimore, MD, 1978), 87, 68.
7 Overall, see L. L. Betzig, 'Medieval monogamy', *Journal of Family History*, 20 (1995), 181–215.
8 E. Westermarck, *History of Human Marriage* (5th edn, London, 1921), iii, 29, 34; see too J. Goody, *Production and Reproduction* (Cambridge, 1973).
9 L. Betzig, 'Sex, succession, and stratification in the first six civilizations: how powerful men reproduced, passed power on to their sons, and used power to defend their wealth, women, and children', in L. Ellis (ed.) *Social Stratification and Socioeconomic Inequality*, vol. 1 (Westport, CT, 1993), 37–74. See too, L. Betzig, *Despotism and Differential Reproduction: a Darwinian view of history* (Hawthorne, NY, 1986) and L. Betzig, 'Politics and sex: the Old Testament case', *Human Nature* (in press). Quotes from S. N. Kramer, *The Sumerians* (Chicago, 1963), 318, and R. van Gulik, *Sexual Life in Ancient China* (London, 1974), 94–5.
10 Hollingsworth, 'Demography of the British peerage', 44, 47, 50.
11 P. Laslett, 'Introduction: comparing illegitimacy over time and between cultures', in P. Laslett, K. Oosterveen and R. Smith (eds) *Bastardy and its Comparative History* (Cambridge, MA, 1980), 18, 20.
12 As Lawrence Stone, a critic of parish register demography, sums up: 'There are many dangers inherent in such projects, the most serious of which is that to some extent the conclusions drawn from these highly costly and labour-intensive quantitative studies still depend on the utility and reliability of the variables selected for study', in L. Stone, *Past and Present Revisited* (London, 1987), 37–8. Among criticisms of the 'illegitimacy ratio' see: Laslett, 'Introduction'; P. Laslett, *Family Life and Illicit Love in Earlier Generations* (Cambridge, 1977), 120–4; P. Laslett, *The World We Have Lost*, 3rd edn (Cambridge, 1984), 155ff; E. Shorter, J. Knodel and E. van de Walle, 'The decline of non marital fertility in Europe, 1850–1960', *Population Studies*, xxv (1971), 380–1; A. MacFarlane, 'Illegitimacy and illegitimates in English history', in Laslett *et al*, *Bastardy*, 78; L. Stone, *Road to Divorce: England* (Oxford, 1990), 104–12. On

misassigned paternity estimates in the 'Liverpool flats' and other spots, see R. Baker and R. Bellis, 'Number of sperm in human ejaculates varies in accordance with sperm competition theory', *Animal Behaviour*, xxxvii (1989), 867–9 and R. Baker and R. Bellis, *Human Sperm Competition* (London, 1995).

13 P. E. H. Hair, 'Bridal pregnancy in rural England in earlier centuries', *Population Studies*, lxvi (1966), 233–43; P. E. H. Hair, 'Bridal pregnancy in earlier rural England further examined', *Population Studies*, xxiv (1970), 59–70. In his book *Before the Bawdy Court: selections from church court and other records relating to the correction of moral offences in England, Scotland and New England, 1300–1800* (London, 1972), 232, Hair concludes: 'The commonest entry in the church court correction records relates to the offense of "ante-nuptial fornication", a charge normally brought against both marriage partners. This usually mean that the wife had been pregnant at marriage and that a child had been born within six months or so thereafter.' Hair suggests that bridal pregnancy rates have risen over time; K. Wrightson and D. Levine, *Poverty and Piety in an English Village: Terling 1525–1700* (Cambridge, 1979), 131, find that while 33 per cent of sixteenth-century brides were with child, only 20 per cent were in the first half of the seventeenth century, and just 11 per cent were in the second half of the seventeenth century.

14 Laslett, 'Introduction', 55.

15 R. M. Smith, 'Marriage processes in the English past: some continuities', in L. Bonfield, R. M. Smith and K. Wrightson (eds) *The World We Have Gained* (Oxford, 1986), 43–99.

16 For Terling: D. Levine and K. Wrightson, 'The social context of illegitimacy in early modern England', in Laslett *et al.*, *Bastardy*, 163; for Lancashire and Essex: K. Wrightson, 'The nadir of English illegitimacy in the seventeenth century', in Laslett *et al.*, *Bastardy*, 187; for Suffolk: Smith, 'Marriage processes in the English past', in L. Bonfield *et al.*, *The World We Have Gained*, 55; for Colyton, Alcester, Aldenham and Hawkshead: K. Oosterveen and R. M. Smith, 'Bastardy and the family reconstitution studies of Colyton, Aldenham, Alcester and Hawkshead', in Laslett *et al.*, *Bastardy*, 112–13; for Banbury and Hartland: S. Stewart, 'Bastardy and the family reconstitution studies of Banbury and Hartland', in Laslett *et al.*, *Bastardy*, 132–3; on Scotland: C. Smout, 'Aspects of sexual behaviour in nineteenth-century Scotland', in Laslett *et al.*, *Bastardy*, 196; and on servants by county, Laslett, 'Introduction', 56.

17 These data contradict Laslett's conclusion that 'the minority of bastardies which arose in situations of social asymmetry, that is from victimization, especially of woman servants, from the keeping of mistresses and from casual encounters between ill-assorted people often strangers to each other, are prominent in the evidence' – when a poor girl had a rich boy on the ropes, she was more likely to take him to court. Laslett, 'Introduction', 56; also Laslett, *World Lost*, 178. On Essex and Lancashire: Wrightson, 'The nadir', 187, and Wrightson and Levine, *Poverty and Piety*, 128.

18 Smout, 'Aspects of sexual-behavior', 194.

19 J. Hajnal, 'European marriage patterns in perspective', in D. Glass and D. Eversley, *Population and History* (London, 1965), 101–43, and J. Hajnal, 'Two kinds of pre-industrial household formation system', in R. Wall, J. Robin and P. Laslett (eds) *Family Forms in Historic Europe* (Cambridge, 1983), 65–70; A. MacFarlane, *Marriage and Love in England 1300–1840* (Oxford, 1986), 337–44. Also, e.g., M. Mitterauer and R. Seider, *The European Family* (Oxford, 1982); R. A. Houlbrooke, *The English Family 1450–1700* (London, 1984); L. Bonfield (ed.) *The World We Have Gained* (Oxford, 1987).

20 See P. Laslett, 'Size and structure of the household in England over three centuries', *Population Studies*, xxiii (1969), 199–223, on means; on variance see p. 208. In Laslett's words, 'Household and family sizes have a known tendency towards positive skewness; that is to say there is a longish tail above the mean.'

21 C. Given-Wilson, *The Royal Household and the King's Affinity* (New Haven, CT, 1986), 22, 278–9; M. Perkins, *The Servant Problem and the Servant in English Literature* (Boston,

MA, 1928), 74; quote is from J. Stowe, *The Survey of London* (London, 1633), ch. vii. Lesser households held progressively lesser numbers. In 1634, the sheriff of Surrey and Essex had 116 servants in livery, but this was unusual; 30 or 40 was common for a sheriff.

22 See summaries in J. Hecht, *The Domestic Servant Class in Eighteenth-century England* (London, 1956), 3–9; M. Perkins, *The Servant Problem and the Servant in English Literature* (Boston, MA, 1928), ch. i; R. Trumbach, *The Rise of the Egalitarian Family* (London, 1978), 124–5; L. Stone, *The Crisis of the Aristocracy: England 1530–1987* (Oxford, 1990), 212–13; L. Stone and J. Stone, *An Open Elite? England 1540–1880* (Oxford, 1984), 425. Quotes from G. Burnet, *History of his Own Time* (Oxford, 1833), vi, 219 and W. Harrison, *Description of Britaine and England* (London, 1577, 1587), ii, iii. Gregory King's table is reprinted in Laslett, *World Lost*, 32–3.

23 Laslett, 'Size and structure', 219, 222; also P. Laslett and R. Wall, *Household and Family in Past Time* (Cambridge, 1972), 152, and P. Laslett, *Family Life and Illicit Love in Earlier Generations* (Cambridge, 1977), 92–3; R. Wall, 'The household: demographic and economic change in England, 1650–1970', in Wall *et al.* (eds) *Family Forms*, 497; R. Smith, 'Some issues concerning families and their property in England 1250–1800', in R. Smith (ed.) *Land, Kinship, and Life Cycle* (Cambridge, 1984), 33–4. Quote from Hajnal, 'Two kinds of pre-industrial household', 93.

24 Laslett, *Family Life and Illicit Love*, 31, 35; A. Kussmaul, *Servants in Husbandry in Early Modern England* (Cambridge, 1981), 70.

25 Kussmaul, *Servants in Husbandry*, 4. According to Kussmaul, the ratio of male to female servants was 171:100 in craftsmen's households, 121:100 in farmers' households, and 107:100 overall. R. Wall, 'The composition of households in a population of six men to ten women: south-east Bruges in 1814', in Wall *et al.* (eds) *Family Forms*, 421; also A. Fauve-Chamoux, 'The importance of women in an urban environment: the example of Rheims household at the beginning of the industrial revolution', in Wall *et al.* (eds) *Family Forms*, 477–8 and R. Mols, *Introduction à la démographie historique de villes d'Europe du XIV' au XVIII' siècle* (London, 1954–6); P. Colquhoun, *A Treatise on Indigence . . . with proposals for ameliorating the condition of the poor* (London, 1806) in Hecht, *Domestic Servant Class*, 34.

26 Husee's letter in M. St Clare Byrne (ed.) *The Lisle Letters* (Chicago, 1981), iv, 27 (see too editor's comments in iii, 17); Weston's letter in ibid., iii, 330; *The Purefoy Letters, 1735–1753* (edited by Leslie Mitchell, New York, 1973), 142. See also Elizabeth Purefoy's letter of 3 March 1739, in which another girl arrives in service with child; when found out, she is sent home – contrary to her own wishes – to her father.

27 On Bedford, G. Thompson, *The Russels of Bloomsbury 1669–1771* (London, 1940), 244–6, 229–30; J. Hanway, *Eight Letters to his Grace – Duke of – on the Custom of Vailsgiving* (London, 1760), letter viii (printed in M. Perkins, p. 114), D. Hume, *On the Populousness of Ancient Nations* (London, 1752), quoted in Hajnal, 'Two kinds of household', 95. Trumbach notes that attitudes were changing in the eighteenth century: 'Most masters might still forbid servants to marry and form families of their own, but some were changing their views under the pressure of the agreement that all men had an equal right to domesticity. . . . If masters after 1750 were being urged that their servants had the right to marry, they had had to be reminded constantly in the previous generation to keep other hands off the maids', in Trumbach, *Egalitarian Family*, 123, 147.

28 I. Ware, *A Complete Body of Architecture* (London, 1756), 346–7, 413; D. Marshall, *The English Domestic Servant in History* (Cornwall, 1949), 14; A. List, *The Two Phases of the Social Evil* (London, 1861), quoted in Smout, 'Sexual behavior in Scotland', 196–7.

29 S. Treggiari, 'Jobs for women', *American Journal of Ancient History*, i (1976), 94.

30 Hecht, *Domestic Servant Class*, 66ff; F. de la Rochefoucauld, *Mélanges sur l'Angleterre* (translated by S. C. Roberts, edited and published as *A Frenchman in England* by J. Marchand, Cambridge, 1933), 25–6.

31 William of Malmesbury, *Chronicle* (London, 1847), 279.
32 G. R. Quaife, *Wanton Wenches and Wayward Wives* (London, 1979), quotes pp. 48, 111–15, 181–2. As Quaife writes, 'the sexual exploitation of an economically less secure female by an economically superior male, even within the labouring class, was a fact of life.' Nice parallels in continental ecclesiastical records include J. Mundy, *Men and Women at Toulouse in the age of the Cathars* (Toronto, 1990) and E. Le Roy Ladurie, *Montaillou* (translated by Barbara Bray, New York, 1979). As Richard Smith says of Ladurie's study, 'Of course, every student of medieval English manor courts or ecclesiastical court act books has found his equivalent of Ladurie's (or more precisely Fournier's) celebrated Pierre Clergue of Montaillou (if not always as picturesque)', in R. Smith, 'Marriage processes in the English past: some continuities', in L. Bonfield *et al.*, *World Lost*, 56.
33 P. E. H. Hair, *Before the Bawdy Court*, 73; L. Stone, *Uncertain Unions and Broken Lives* (Oxford, 1995), 105, 113–15.
34 Ibid. There are, of course, medieval precedents for all this. Among the many payments that defined a lord's rights, the most definitive may have been *merchet*. Having to pay to give one's daughter in marriage was an indelible mark of subordinate status. Merchets arguably bought *ius primae noctis* off. There has been a long debate about the *ius primae noctis* – the 'right of the first night'. Was it, or was it not, the lord's privilege to spend his wedding night with his tenant's wife? In F. W. Maitland's opinion, 'among the thousands of entries in English documents relating to this payment, it would we believe be utterly impossible to find one which gave any sanction to the tales of a *ius primae noctis*.' Most modern scholars agree. Nevertheless merchet was real; and the man who paid it on his daughter's marriage was legally unfree with respect to the man who got paid. See F. Pollock and F. W. Maitland, *The History of English Law before the Time of Edward I* (edited by S. F. C. Milsom, Cambridge, 1968), ii.i.7, 12. On *ius primae noctis* see, e.g., W. D. Hovarth, 'Droit de Seigneur: fact or fantasy', *Journal of European Studies* (1971), 291–312; E. Searle, 'Seigneurial control of women's marriage: the antecedents and function of merchet in England', *Past and Present*, lxxxii (1979), 3–43; R. Faith, 'Debate: seigneral control of women's marriage', *Past and Present*, xc (1983), 3–43.
35 J. Evelyn, *Diary* (edited by G. de la Bédoyere, 1994), 437 (26 May 1703).
36 *The Diary of Samuel Pepys, M.A., F.R.S.* (edited by R. C. Latham and W. Matthews, London, 1970–83); John Wilson, *Private Life of Mr. Pepys* (New York, 1959), 145, 228.
37 Boswell's affairs are summarised in Stone, *Family, Sex and Marriage*, 575.
38 *Boswell on the Grand Tour* (edited by F. A. Pottle, Melbourne, 1953), 213, 247–8. Lawrence Stone, who reviewed six rich diaries from early-modern England – by Simon Forman, the sixteenth-century necromancer; Robert Hooke, the seventeenth-century inventor; William Byrd, the eighteenth-century gentleman farmer; Sylas Neville, the eighteenth-century man-about-London; Boswell and Pepys – summed up: Women were available to a man of means. Starting at the top, one might find a respectable married woman; failing that, there were actresses, seamstresses, and whores. But, most of all, there were 'the poor amateurs, the ubiquitous maids, waiting on masters and guests in lodgings, in the home, in inns; young girls whose virtue was always uncertain and was constantly under attack.' They were 'the most exploited, and most defenceless, of the various kinds of women whose sexual services might be obtained by a man of quality.' Stone, *Family, Sex, and Marriage*, 601.
39 Stone, *Uncertain Unions and Broken Lives*, 258–96; L. Blanch, *The Game of Hearts: Harriette Wilson's Memoirs* (New York, 1955), 21; H. M. Hyde, *A Tangled Web: sex scandals in British politics and society* (London, 1986), 109; J. Clelland, *Memoirs of a Woman of Pleasure* (reprint New York, 1990), quote p. 276. Blanch perceptively writes of the decline of the courtesan: 'Today, in America, the courtesan may be said to have been replaced by the psychoanalyst. In place of the alcove there is the analyst's office. But basically the functions of both courtesan and analyst have the same principle. Both offer

90   Laura Betzig

escape, relaxation and individual attention; both are expensive. And the couch is still there. *Plus ça change.*'

40 T. Seaton, *The Conduct of Servants in the Great Families* (reprinted London, 1985), 143–5; E. Haywood, *A Present for a Servant-maid* (reprinted London, 1985), 45–6.

41 J. Swift, 'Directions to servants', in *The Works of Jonathan Swift* (edited by Roscoe Thomas, 1861, New York), 136; F. G. Emmison, *Elizabethan Life* (Chelmsford, 1973), 7, 28; H. Walpole, *Correspondence*, xxii, 211 and xxxiii, 149; and W. Acton, *The Functions and Disorders of the Reproductive System* (London, 1858), 207, 200, 129 – both quoted in Stone, *Family, Sex, and Marriage*, 646–7, 519.

42 J. Boswell, *The Life of Samuel Johnson*, 3rd edn (London, 1799; reissued Oxford, 1946), 265, 372; J. Boswell, *The Ominous Years* (edited by C. Ryskamp and F. A. Pottle, New York, 1963), 74, 82; *Boswell's London Journal* (edited by F. A. Pottle, Edinburgh, 1991), 54; *Boswell in Extremes* (edited by C. Weis and F. A. Pottle, New York, 1959), 28; J. L. Clifford, *Hester Lynch Piozzi (Mrs. Thrale)* (Oxford, 1968), 98; K. C. Balderston (ed.) *Thraliana* (Oxford, 1942), ii, 967.

43 T. C. Phillips (Muilman), *A Letter Humbly Address'd to the Right Honourable the Earl of Chesterfield* (London, 1750), 14; Chesterfield, Philip Dormer Stanhope, *Letters* (edited by B. Dobrée, London, 1932); R. Addison and J. Steele, *Selections from the Tatler and the Spectator* (edited by A. Ross, Harmondsworth, 1982), 263, 265, 273; D. Defoe, *Conjugal Lewdness, or, Matrimonial Whoredom* (orig. London 1727, reprint edited by M. Novak, Gainesville, FL, 1967), 23.

44 In A. N. Jeffares (ed.) *Restoration Comedy* (London, 1974), vol. 1, 55, 267, 465; vol. 4, 217.

45 Jeffares, *Restoration Comedy*, vol. 1, quotes on xv, 320, 335, 350, 355, 361.

46 D. Defoe, *Robinson Crusoe* (New York, 1991), 101. When they finally found him, his tobacco was in his pocket and he was smoking his knife (D. Singleton, personal communication).

47 D. Defoe, *Moll Flanders* (New York, 1989), 9, 13, 15, 39, 41.

48 S. Richardson, *Pamela* (Harmondsworth, 1985), 49, 143, 145, 173, 217, 246, 422, 473. Fielding sums the plot up in *Shamela*: 'Hussey, he says, don't provoke me, I say. You are absolutely in my power, and if you won't let me lie with you by fair Means, I will by Force' (New York, 1987), 291.

49 S. Richardson, *Clarissa* (edited by A. Ross, Harmondsworth, 1985), 50, 213, 417, 419, 669–70. Lovelace is an advocate of annual marriages. 'Determined to marry I would be, were it not for this consideration: that once married, and I am married for life. That's the plague of it! – Could a man do as the birds do, change every Valentine's day (a *natural* appointment! for birds have not the *sense*, forsooth, to fetter themselves, as we wiseacre men take great and solemn pains to do); there would be nothing at all in it' (p. 872).

50 H. Fielding, *The History of Tom Jones* (edited by P. R. C. Mutter, Harmondsworth, 1985), 170, 178, 182, 213–14.

51 I. Watt, *The Rise of the Novel* (Berkeley, CA, 1967), quotes on 53, 58. Watt's comments on Richardson are particularly prescient. He writes on p. 47: 'Pamela, then, may be regarded as the culture-heroine of a very powerful sisterhood of literate and leisured waiting-maids. We note that her main stipulation for the new post she envisaged taking up after leaving Mr. B. was that it should allow her "a little time for reading." This emphasis prefigured her triumph when, following a way of life rare in the class of the poor in general but less so in her particular vocation, she stormed the barriers of society and of literature alike by her skilful employment of what may be called conspicuous literacy' – that is, by writing letters.

52 Quaife, *Wanton Wenches*, 134, 139. Other times, adulterers used force. One Glastonbury wife watched while her lover 'hath beaten her husband . . . and she hath upholden him in it.' Two other men hid while their wives were raped; they 'would not for fear of their lives come forth' to defend their women's honour.

53 R. Trumbach (ed.) *Trials for Adultery* (New York, 1985), vols i–vii, quotes from iii, 1–4 (Arnold), iii, 1–3 (Robinson), iii, 1–2 (Oliver).

54 Ibid., vol. vii (Cibber); H. M. Hyde, *A Tangled Web: sex scandals in British politics and society* (London, 1986), 68–73, 111–40 (Norton, Dilke); Stone, *Uncertain Unions and Broken Lives*, 567 (Westmeath); A. Hamilton, *Count Gramont at the Court of Charles II* (edited and translated by N. Deakin, London, 1965), 80.

55 As Stone, *Family, Sex, and Marriage*, 554–5, writes: 'Almost without exception he confined actual sexual intercourse to married women whose husbands were not too long absent, so that if pregnancy occurred he could not be held responsible.' Also Wilson, *Private Life of Mr. Pepys*, who says on p. 99: 'The wives and daughters of the poor were fair game for any gentleman. The humble cuckold had no "honor" to lose, and could only hope to profit from his wife's shame.'

56 Pepys, *Diary*, 10 November 1668.

57 *Boswell on the Grand Tour*, 248.

58 D. Defoe, *Roxana* (Oxford, 1981), 8, 28, 65, 70, 79–81, 172.

50 H. Fielding, *Amelia* (edited by D. Blewett, Harmondsworth, 1987), 226–7, 300, 308, 544. As one critic, Harrison Steeves, points out, 'The sexual manners of the British aristocracy at that time can well be called bad manners.' As he also makes clear, times were changing. 'Within another half-century the rake was to have gone out of vogue as an agreeable, or even an amusing, figure, and young men had taken their place in fiction who would not have recognized a wild oat.' H. R. Steeves, *Before Jane Austen* (New York, 1965), 89, 170.

60 Jeffares, *Restoration Comedy*, vol. 2, 16, 460, 549–50; vol. 3, 309.

61 Jeffares, *Restoration Comedy*, vol. 2, 362; vol. 4, 94, 287.

62 T. Killigrew, 'The Parson's Wedding', in Jeffares, *Restoration Comedy*, vol. 1, 56.

63 See O. Mandel, *The Theatre of Don Juan: a collection of plays and views, 1630–1963* (Lincoln, NE, 1963). Quotes from his translation of Molina are on 70, 78, 95, 97.

64 Quote in Mandel, *Theatre of Don Juan*, 445.

65 Quotes in Mandel, *Theatre of Don Juan*, 123, 292.

66 Byron, *Don Juan* (edited by T. Steffan, E. Steffan and W. Pratt, Harmondsworth, 1982), quotes from Cantos ii, 213; iii, 8; ix, 55–6; xi, 77.

67 Quotes are from J. Goethe, *Faust* (translated by Walter Kaufmann, New York, 1990), part I, lines 358–9, 1590, 1635–9. A few more telling lines from Faust: 'I loathe the knowledge I once sought./In sensuality's abysmal land/Let our passions drink their fill!' (part I, lines 1749–51) and 'One glance from you, one word gives far more pleasure/Than all the wisdom in the world' (part I, lines 3079–80) and, not least significant, 'Oh, that is fine, for it's unpleasant/To visit her without a present' (part I, lines 3674–5). Quote from Nietzsche comes from Walter Kaufmann's introduction to this *Faust* edition, p. 7.

68 G. B. Shaw, *Man and Superman* (edited by Dan Laurence, Harmondsworth, 1957), all quotes from Act 3; A. Camus, *The Myth of Sisyphus* (translated by Justin O'Brien, London, 1955), 11, 62; M. Foucalt, *The History of Sexuality*, vol. 1 (translated by Robert Hurley, New York, 1990), 3, 11, 39–40. Literary critics had already established Don Juan's impotence, narcissism and love for his own sex; see e.g. essays in Mandel, *Theatre of Don Juan*. See too S. Kierkegaard, *Either/Or* (translated by H. V. Hong and E. H. Hong, Princeton, NJ, 1987), 49, 99, 104. Kierkegaard (to whom music is the apotheosis of life, Mozart the apotheosis of musicians, and *Don Giovanni* the apotheosis of Mozart – 'Immortal Mozart! You to whom I owe everything – to whom I owe that I lost my mind') admits it might be hard to find a young man unwilling 'to have given half his kingdom to be a Don Juan, or perhaps all of it.' But, being a philosopher, he seems to conclude that Juan embodies the 'sensual' *idea* of his time – anticipated by 'psychical' Greeks like Ulysses, and followed by 'intellectual' moderns like Faust. Kierkegaard finds 'something very profound' in the fact that Faust gets just one girl, while Juan gets more or less any girl; but he's not sure what.

69 V. Gibbs, *Complete Peerage* (London, 1926); C. Given-Wilson and A. Curteis, *The Royal Bastards of Medieval England* (London, 1984); and K. MacDonald, 'The establishment and maintenance of socially imposed monogamy in Western Europe', *Politics and the Life Sciences*, xiv (1995), 3–23. Together these two lists cover kings from William the Conqueror (1066) to George VI (1950). None of these numbers comes close to the European record, held by Augustus the Strong, Elector of Saxony and King of Poland in the eighteenth century, who had no less than 354 acknowledged bastards – cited in L. Stone, *Family, Sex, and Marriage*, 532–3. The differences between Augustus the Strong, other exalted Europeans, and English kings may lie in the numbers of bastards produced, the proportion of bastards acknowledged, or both.

70 C. Newman, 'Review essay of "Royal bastards",' *Medieval Prosopography*, 7 (1986), 91; Byrne, *Lisle Letters*, 4, 14. Interestingly, in dating the year of Lisle's birth, Byrne uses a tailor's bill of 1472, listing items for 'my Lord the Bastard', sent by Edward IV to the Exchequer for payment. That makes Lisle, at many events in his life, quite a bit older than his peers. Byrne says 'it is perhaps unwarrantable to assume that he was the King's only bastard son, but no record of any other has survived', p. 139.

71 On Henry I's empire building: W. C. Hollister and T. F. Keefe, 'The making of the Angevin empire', *Journal of British Studies*, xii (1973), 5.

72 For 'identifiable' bastards only, $t = 2.276$, d.f. $= 19$, $p = 0.0152$; for 'identifiable', 'possible' plus 'doubtful' bastards, $t = 2.300$, d.f. $= 19$, $p = 0.0144$ (see Table 3.1). Edward III, who was survived by three legitimate sons but succeeded by a grandson, is included among kings succeeded by sons; royal consorts – Mary I's husband Philip II, Mary II's husband William III, and Victoria's husband Albert – are not included in either group, nor are Edward V and VI (both of whom died in their minorities), nor is Edward VIII (who abdicated), nor is Cromwell (who was not a king). Quote is from Given-Wilson and Curteis, *Royal Bastards*, 177.

73 Fisher's Exact Test, $p = 0.084$ (see Table 3.2). Again neither royal consorts (Mary I's husband Philip II), nor kings who died in their minority (Edward V and Edward VI), are included here. Quote is from William of Malmesbury, *Chronicle of the Kings of England* (translated by J. A. Giles, London, 1847), 447. On 'licentiousness' as libel, see e.g. Arnaldo Momigliano, *The Development of Greek Biography* (Cambridge, MA, 1971) and Ronald Syme, *The Roman Revolution* (Oxford, 1939).

74 Of earlier kings' chroniclers are sketchy but consistent. Gildas, who hated everybody, hated royalty. He said in brief: 'Britain has kings, but they are tyrants. . . . They often plunder and terrorize – the innocent; they defend and protect – the guilty and thieving; have many wives – whores and adulteresses.' He went on in detail about 'the five tyrants'. Constantine, he said, put away his wife and wallowed in the 'stench' of adultery; Aurelius Caninus was a killer, fornicator and adulterer; Vortipor was similar; Cuneglasus, Latin for 'red butcher', loved his widow's sister – 'aside from countless other lapses'; Maglocunus, 'last in my list, but first in evil', was the most extravagant sinner. The *Anglo-Saxon Chronicle* mentions, in passing, Cynewulf's mistress and Aethelwold's abduction of a nun. William of Malmesbury disliked Vortigern, a man 'wholly given up to the lusts of the flesh, the slave of every vice: a character of insatiable avarice, ungovernable pride, and polluted by his lusts'; Edwy was 'a wanton youth, who abused the beauty of his person in illicit intercourse'; Edgar, travelling in Andover, 'ordered the daughter of a certain nobleman, the fame of whose beauty had been loudly extolled, to be brought to him.' And Geoffrey of Monmouth remembers Ebraucus, in the time of King David, who was 'very tall and a man of remarkable strength', made war on the Gauls, founded 'the castle of Mount Agned, which is now called the Maidens' Castle' and 'by the twenty wives which he had, he was the father of twenty sons and of thirty daughters.' In Gildas, *The Ruin of Britain* (translated by M. Winterbottom, Chichester, 1978), 29–32; *Anglo-Saxon Chronicle*, 755[757], 901[899] (translated by G. N. Garmonsway, New York, 1967); William of Malmesbury, *Chronicle of the Kings of England* (translated by J. A. Giles, London, 1847),

7, 145, 160; Geoffrey of Monmouth, *History of the Kings of Britain* (translated by Lewis Thorpe, Harmondsworth, 1966), ii, 7–8. See too M. C. Ross, 'Concubinage in Anglo-Saxon England', *Past and Present*, cviii (1985), 3–34.

75 D. C. Douglas, *William the Conqueror* (Berkeley, CA, 1964). Quotes from *Anglo-Saxon Chronicle*, D, 1087; Orderic Vitalis, *Ecclesiastical History* (translated by M. Chibnal, Oxford, 1980), vii, 16; William of Jumièges, *Deeds of the Norman Dukes*, AD 1087 (translated by Elisabeth van Houts, Oxford, 1995); William of Poitiers, *Deeds of William*, i, 48 (translated by R. H. C. Davis and M. Chibnall, Oxford, 1998); William of Malmesbury, *Chronicle*, 305.

76 F. Barlow, *William Rufus* (London, 1983). Quotes from Eadmer, *Historia Novorum* (edited and translated by G. Bosanquet and R. W. Southern, London, 1964), 192–3; Orderic Vitalis, *Ecclesiastical History*, iv, 8 (vol. 4, book 8), v, 10; William of Malmsbury, *Chronicle*, 337. Barlow suggests that, as Rufus left no heirs, and no one thought him impotent, 'it is likely that he was either sterile or homosexual' (p. 109); see Malmesbury, *Chronicle*, 428.

77 Hollister and Keefe, 'Angevin empire'; Orderic Vitalis, *Ecclesiastical History*, vol. 5, book x, p. 299; vol. 6, book xi, p. 99; William of Malmesbury, *Chronicle of the Kings of England* (translated by J. A. Giles, London, 1847), 7, 145, 160, 216, 305, 337, 428, 447; Henry of Huntington, *Chronicle* (translated by T. Forester, New York, 1853), book viii, 261–2, 335; *Gesta Stephani* (edited and translated by K. R. Potter and R. H. C. Davis, Oxford, 1976), i.

78 R. H. C. Davis, *King Stephen* (3rd edn, New York, 1990). Quotes from *Anglo-Saxon Chronicle* (translated by G. N. Garmonsway, New York, 1967) AD 1137; Robert de Torigni, *Chronica* (translated by J. Stevenson, London, 1991) AD 1140; *Gesta Stephani*, xii.

79 W. L. Warren, *Henry II* (London, 1973); A. Kelly, *Eleanor of Aquitaine* (Cambridge, MA, 1950). Quotes on Eudo de Porhoet and Roger of Clare in Given-Wilson and Curteis (1986), 98; Gerald of Wales, *Autobiography* (translated by H. E. Butler, London, 1937), i, 8; Roger of Hoveden, *Annals* (translated by H. T. Riley, London, 1853) AD 1191; Gerald of Wales, *Journey* (translated by L. Thorpe, Harmondsworth, 1978) p. 45; Ralph Niger, *Chronicles* (edited by R. Anstruther, New York), 168; Walter Map, *Courtier's Trifles*, v, 6 ('sober . . . dreams'); *Guernes de Pont-Sainte-Maxence*, vv 336f, 416f, translated in F. Barlow, *Thomas Becket* (London, 1986).

80 J. B. Gillingham, *Richard the Lionheart* (2nd edn, London, 1989) and J. B. Gillingham, *Richard, Coeur de Lion* (London, 1994), 135. On his homosexuality, J. Brundage, *Richard Lion Heart* (London, 1974). Other quotes from Roger of Howden, *The History of England and of Other Countries of Europe from AD 732 to AD 1201* (translated by H. T. Riley, London, 1853), AD 1195; Imad ad-Din cited in Gillingham (1989), 184; Walter of Guisborough, *Chronicle* (edited by H. Rothwell, London, 1957), 142. On bisexuality in Greece, K. Dover, *Greek Homosexuality* (London, 1996); in Rome, L. Betzig 'Roman polygyny'; among later kings of England see p. 75.

81 W. L. Warren, *King John* (Berkeley, CA, 1978); S. Painter, *The Reign of King John* (Baltimore, MD, 1949), 231–4.

82 D. A. Carpenter, *The Minority of Henry III* (London, 1990). Quotes from Matthew Paris, *English History* (translated by J. A. Giles, London, 1852) AD 1239, 1243; *Song of Lewes*, 1, 155–6 (in T. Wright, *Political Songs*, London, 1996); *Chronicle of Bury St Edmonds*, AD 1265 (translated by A. Grandsen, London, 1964); *Provisions of Oxford*, 1258 (translated in H. Rothwell, *English Historical Documents*, III, 361–7, London, 1975).

83 M. Prestwich, *Edward I* (Berkeley, CA, 1988), 129. Quotes from Pierre de Langtoft, *Chronicle*, AD 1296 (translated by T. Wright, London, 1964); *Song of Lewes*, lines 443–4; *Chronicle of Bury St Edmunds*, AD 1269; Thomas Wykes, *Chronicle*, iv, 264 (edited by H. Luard, London, 1965); J. H. Round, *King's Serjeants* (London, 1911), 97 on whore-keepers; *Statute of Westminster* I, 1275, c.32, translated in H. Rothwell, *English Historical Documents* (London, 1975), iii, 397–409. See B. Byerly and C. Byerly, *Records of the*

Wardrobe and Household 1285–86 (London, 1977) on Edward's household, and Prestwich (1988), 159 on Christmas.

84   H. F. Hutchinson, *Edward II: The Pliant King* (London, 1971). Quotes from Froissart, *Chronicles*, AD 1327 (translated by G. Bereton, Harmondsworth, 1968); articles of accusation in *Foedera*, ii, 650 (T. Rymer, London, 1816–33); Walter of Guisborough, *Chronicle*, AD 1307 (edited by H. Rothwell, London, 1957); *Vita Edwardi Secundi Monachi*, AD 1307 (translated by N. Denholm-Young, London, 1957); *New Ordinances*, 1311, *c*.8, 10 (translated in H. Rothwell, *English Historical Documents*, iii, 527–39, London, 1975); *Household Ordinances*, 1318 (translated in part in C. Stephenson, *Sources of English Constitutional History*, New York, 1937, 199–204). See Round 1911, 97, on court whores.

85   W. M. Ormrod, *The Reign of Edward III* (New Haven, CT, 1990). Quotes from Froissart, *Chronicles*, AD 1327; William Langland, *Piers Plowman*, iii, 122–6 (translated by Frank Goodrich, Harmondsworth, 1959); *Brut*, ii, 333 (edited by F. Brie, London, 1906); R. Allen Brown, *English Castles* (London, 1976), 208 on 'Versailles'; *De Speculo Regis* (edited by J. Moisant, Paris, 1891), 103; *Rotuli Parliamentarum*, ii, 268–70 (London, 1767–83). See M. Prestwich, *War, Politics, and Finance under Edward I* (London, 1972), J. Maddicott, *English Peasantry and the Demands of the Crown* (London, 1975) and C. Given-Wilson, *Royal Household and the King's Affinity* (New Haven, CT, 1986) on supplying the king.

86   J. L. Kirby, *Henry IV of England* (London, 1970), A. Tuck, *Richard II and the English Nobility* (London, 1973), C. Allmand, *Henry V* (London, 1992), R. A. Griffiths, *Reign of King Henry VI* (London, 1981). Articles of desposition are in *Rotuli parliamentarum*, iii, 417–53. Other quotes from Henry Knighton, *Chronicle*, AD 1380 (translated by G. H. Martin, Oxford, 1995); *Westminster Chronicle*, AD 1387 (translated by L. C. Hector and B. F. Harvey, London, 1982); Thomas Walsingham, *Chronica majora*, AD 1397 (translated in part in C. Given-Wilson, *Chronicles of the Revolution*, New York, 1993); *Vita ricardi secundi*, AD 1397 (translated in part in C. Given-Wilson, *Chronicles of the Revolution*, New York, 1993). See Given-Wilson 1986, 94, on £53,200; see R. A. Brown *et al.*, *History of the King's Works* (London, 1963), ii, 934 on 2,000 tiles. Thomas Haxey's petition is in *Rot parl.*, iii, 338.

87   Shakespeare, *Henry IV Part 1*, i, 2; *Henry IV Part 2*, ii, 4; E. Hall, *The Vnion of the Two Noble and Illustre Famelies of Lanastre & Yorke . . .* (New York, 1965), 113, 303; R. Holinshed, *Chronicles* (London, 1807), iii, 57, 133, 324; *Deeds of Henry V*, 22 (translated by F. Taylor and J. Roskell, Oxford, 1975).

88   C. D. Ross, *Edward IV* (London, 1974). Quotes from Dominic Mancini, *The Usurpation of Richard the Third* (translated by C. Armstrong, Oxford, 1984), 59, 61, 67; Philip Comines, *Memoirs* (edited and translated by S. Kisner and I. Cazeaux, Columbia SC, 1969), vi, 12; Thomas More, *The History of Richard III* (in R. S. Sylvester (ed.) *The Complete Works of Thomas More*, New Haven, CT, 1963), 5, 55, 69, 72; Polydore Vergil, *English History*, 117, 172; *Black Book of Edward IV*, in A. R. Myers, *Household of Edward IV* (Manchester, 1959); and Stubbs 1897, ii, 21 on magnates' expenses; John Fortescue, *In Praise of the Laws of England* (translated by S. Lockwood, Cambridge, 1997), 35–6.

89   C. D. Ross, *Richard III* (London, 1981) and R. Horrox, *Richard III: a study of service* (New York, 1989). Quotes from Polydore Vergil, 190; More, *History*, 6, 8; Holinshed, *Chronicles*; Hall, *Vnion*, 420; More, *History*, 170–1; *Crowland Chronicle* (edited by N. Pronay and J. Cox, London, 1986), 572.

90   S. B. Chrimes, *Henry VII* (London, 1972). Quotes from Polydore Vergil, *History*, 26; Desiderius Erasmus, *Complete Works: Poems* (translated by C. H. Miller, Toronto, 1993), 4; Francis Bacon, *History of the Reign of King Henry VII* (edited by Brian Vickers, 1998), 68, 203; John Fisher, *English Works* (edited by J. Maynor, London, 1935), i, 269.

91   J. J. Scarisbrick, *Henry VIII* (Berkeley, CA, 1968); H. Miller, *Henry VIII and the English Nobility* (Oxford, 1986). Quotes from Hall, *Vnion*, 703; Holinshed, *Chronicles*, iii, 863;

H. Savage, *The Love Letters of Henry VIII* (London, 1949), 28, 39, 69, 93; *Letters and Papers of the Reign of Henry VIII*, i, 285, i, 1349, ii, 3437, viii, 567, x, 908, xii, 2.764, xv, 822, 823, xvi, 106 (ed. J. Gairdner, London, 1862–1910); Thomas More, *Utopia* (edited by E. Surtz, New Haven, CT, 1964), 47. See J. Goody, *The Development of the Family and Marriage in Europe* (Cambridge, 1983), 185, for insights on Henry VIII's succession.

92 For example G. W. O. Woodward, *The Dissolution of the Monasteries* (London, 1966); C. G. A. Clay, *Economic Expansion and Social Change in England 1550–1700* (Cambridge, 1984); D. C. Coleman, *The Economy of England 1450–1750* (Oxford, 1977); F. C. Dietz, *English Public Finance* (London, 1964); R. Floud and D. McCloskey, *The Economic History of Britain since 1700* (Cambridge, 1994); J. Thirsk, *The Agrarian History of England, 1500–1640* (Cambridge, 1967); L. Stone, 'The educational revolution in England, 1560–1640', *Past and Present*, xxviii (1964), 41–80, and L. Stone, 'Literacy and education in England, 1640–1900', *Past and Present*, xlii (1969), 69–139; E. A. Wrigley and R. S. Schofield, *The Population History of England 1541–1871* (Cambridge, 1981); W. MacCaffrey, *Elizabeth I* (London, 1993); K. Sharpe, *The Personal Rule of Charles I* (New Haven, CT, 1992); G. R. Elton, *The Parliament of England 1559–1581* (Cambridge, 1986); C. Hill, *Century of Revolution 1603–1714* (New York, 1980). Quotes are from J. P. Sommerville (ed.) *King James VI and I Political Writings* (Cambridge, 1994), 181, and S. J. Houston, *James I* (London, 1973), 37, 123.

93 *Constitutio Domus Regis* (edited and translated by Charles Johnson with F. Carter and D. Greenway, Oxford, 1983) and G. H. White, 'The household of the Norman kings', *TRHS*, 4th series, xxx (1948), 127–55; Walter Map, *De nugis curialum* (translated by M. R. James, C. Brooke and R. Mynors, Oxford, 1983), i, 10; T. F. Tout, *Chapters in Medieval Administrative History* (London, 1920), ii, 59; *Records of the Wardrobe and Household* (edited by B. and C. R. Byerly, London, 1977 and 1986); Michael Prestwich, *Edward I* (London, 1988); Rosemary Horrox, *Richard II: a study of service* (New York, 1989), 232; A. R. Myers, *The Household of Edward IV: the black book and the ordinance of 1478* (Manchester, 1959); C. Given-Wilson, *The Royal Household and the King's Affinity* (New Haven, CT, 1986), quotes on purveyors p. 42.

94 S. J. Houston, *James I* (London, 1973), 25; F. Dietz, *English Public Finance*, 107, 166, 270; G. E. Aylmer, *The King's Servants: the civil service of Charles I* (London, 1974), 27; J. M. Beattie, *The English Court in the Reign of George I* (Cambridge, 1967), 279–82; C. Hibbert, *George IV: Prince of Wales* (London, 1972), 176; C. Hibbert, *George IV: regent and king* (London, 1973), 83; R. Ingpen, *The Letters of Percy Bysshe Shelley* (London, 1909), i, 99–100; C. Hibbert, *Edward VII: a portrait* (London, 1976), 65, 97, 233.

95 Houston, *James I*. Quotes from Sommerville, *King James VI and I Political Writings*, 23, 38–39, 41; J. Oglander, *A Royalists's Notebook* (London, 1936); E. Peyton, *The Divine Catastrope of the Kingly Famioly of the House of Stuarts* (in W. Scott (ed.) *Secret History of the Court of James I*, Edinburgh, 1811), 353; A. Weldon, *Character of King James* (London, 1807), 56; S. D'Ewes, *Autobiography* (edited by J. O. Halliwell, London, 1845), 170; L. A. Hutchinson, *Memoirs of the Life of Colonel Hutchinson* (edited by J. Sutherland, New York, 1994), 42–6; F. Osborne, *Traditionall Memoyres of the Raigne of King James the First* (in Sir W. Scott (ed.) *Secret History*, Edinburgh, 1811), 217–18; J. Fletcher's play reprinted in F. Bowers (ed.) *The Dramatic Works in the Beaumont and Fletcher Canon* (Cambridge, 1982), II.iii.

96 K. Sharpe, *The Personal Rule of Charles I* (New Haven, CT, 1992). On Charles's diet and exercise, see T. Herbert, *Memoirs of the Two Last Years of Charles I* (London, 1813); quotes from L. Hutchinson, *Memoirs*, i, 42–6; E. Hyde, *The History of the Rebellion and Civil Wars in England* (edited by W. D. Macray, Oxford, 1988), xi, 238–9, 243; R. Reade in *Calendar of State Paperos Domestic, 1639–40*, 365.

97 R. Hutton, *Charles the Second: King of England, Scotland and Ireland* (Oxford, 1989); A. I. Dascent, *The Private Life of Charles II* (London, 1927); J. H. Wilson, *All the King's Ladies* (London, 1958). Quotes from J. H. Wilson (ed.) *Court Satires of the Restoration* (Columbus, OH, 1976), 63; A. Marvell, 'The Kings Vowes', reprinted in H. M. Margoliouth (ed.)

The Poems and Letters of Andrew Marvell (Oxford, 1971), lines 37–46; A. Hamilton, *Memoirs of the Court of Charles the Second by Count Gramont* (edited by W. Scott, London, 1859), 105, 149–50, 173, 259, 295; G. Burnett, *History of his Own Time* (London, 1724), i, 93–4; J. Evelyn, *Diary*, 4 October 1683 and 6 February 1685.

 98 J. Miller, *James II: a study in kingship* (Sussex, 1977). Quotes from Hamilton, *Count Gramont*, pp. 59, 113, 274, 276; Burnet, *History*, i, 623. Anne Hyde was thought by many, including Anthony Hamilton, a bad match: 'It was farcical to advance the daughter of an insignificant lawyer, who had been raised by the King's favour to the peerage without being of noble blood, and to the office of Chancellor without any abilities.'

 99 R. Hatton, *George I* (Cambridge, MA, 1978); quotes from H. Walpole, *Memoirs of King George II* (edited by J. Brooke, New Haven, CT, 1985), iii, 122.

100 C. Trench, *George II* (London, 1973); quotes from H. Walpole, *Memoirs of King George II*, i, 116–17; J. Hervey, *Some Materials towards Memoirs of King George II* (edited by R. Sedgwick, London, 1931), 43, 382, 457–8, 491, 538–9, 559, 744–5.

101 S. E. Ayling, *George the Third* (London, 1972); quotes from A. Aspinall (ed.) *Correspondence of George, Prince of Wales, 1770–1812* (London, 1968), i, 34; v, 114, 117; M. Delany, *Autobiography and Correspondence* (edited by Lady Llandover, London, 1862), ii, 474; Byron's 'Vision of judgment', lines 359–62, reprinted in D. Perkins (ed.) *English Romantic Writers* (New York, 1967).

102 C. Hibbert, *George IV: Prince of Wales* (London, 1972) and C. Hibbert, *George IV: regent and king* (London, 1973); quotes from D. Lieven, *Private Letters of Princess Lieven to Prince Metternich 1820–1826* (edited by P. Quennell, London, 1948), 182; C. Granville (ed.) *Lord Granville Leveson Gower (First Earl Granville) Private Correspondence 1781–1821* (London, 1916), ii, 297; B. Haydon, *Diary* (edited by W. B. Page, Cambridge, MA, 1960), ii, 350.

103 T. Pocock, *Sailor King: the life of King William IV* (London, 1991), 66, 91, 122, quote Royal Archive letters. Other quotes from A. Aspinall (ed.) *The Later Correspondence of George III* (Cambridge, 1962), i, 268; W. Dyott, *Diary* (London, 1962), 36–45.

104 C. Hibbert, *Edward VII: a portrait* (London, 1976), 47–8, 73–4, 150, 244; T. Aronson, *The King in Love: Edward VIII's mistresses* (London, 1988).

105 J. Gore, *King George V: a personal memoir* (London, 1941), quotes pp. 11, 116; letter from Duke to Duchess of York excerpted in J. Pope-Hennessy, *Queen Mary* (London, 1959), 280.

106 K. Rose, *King George V* (London, 1983), 90–6, 200, 326–7.

107 S. Bradford, *King George VI* (London, 1989), 112, 165–6.

108 On sex in the Bible, councils and canon law, the authority is J. Brundage, *Law, Sex, and Christian Society in Medieval Europe* (Chicago, 1987). On polygyny in Rome: Betzig, 'Roman polygyny'; on polygyny in the Middle Ages: Betzig, 'Medieval monogamy'; on polygyny in other civilizations: Betzig, *Despotism*, and Betzig, 'Sex and succession'.

109 J. S. Mill, *The Subjection of Women* (original 1869, reprint Indianapolis, IN, 1988), 100; M. Wollstonecraft (original 1792, reprint New York, 1982), 73; Cicero, *Letters* (Harmondsworth, 1990), dated 29 April, 5 October and 29 November, 58 BC. Marital respect goes back even farther, of course, than Cicero; it seems to go back as far as marriage. On, for instance, a fourteenth-century BC titulary of Nefertity from Karnak in Egypt, the text reads: 'Heiress, Great of Favor, Mistress of Sweetness, beloved one, Mistress of Upper and Lower Egypt, Great King's Wife, whom he loves, Lady of the Two Lands, Nefertity' – though her husband, Akhenaten, had hundreds of girls in his harem (reprinted in D. Redford, *Akhenaten: the heretic king*, Toronto, 1984), 133. In Tudor England, it was not new to find Arthur Lisle writing his wife: 'Mine own sweetheart, In the heartiest manner that I can I commend me unto you, never thinking so long for you as I do now; trusting that you will be here shortly', or to find Honor Lisle answer: 'Even with my whole heart root I have me most heartily recommended unto you . . . for I can neither sleep nor eat nor drink that

doth me good, my heart is so heavy and full of sorrow, which I know well will never be lightened till I be with you, which I trust shall be shortly' (letters dated 2 and 3 December 1538, in M. Byrne, *The Lisle Letters*, vol. 5, 318–19). Compare A. MacFarlane, *Marriage and Love in England* (Oxford, 1986) and *The Family Life of Ralph Josselin* (Cambridge, 1970). On marriage and money see, e.g., Westermarck, *History Marriage*; G. P. Murdock, *Social Structure* (New York, 1949); J. Goody, *Production and Reproduction* (Cambridge, 1976).

110 Outstanding reviews of dominance and reproduction in other organisms include T. H. Clutton-Brock (ed.) *Reproductive Success* (Cambridge, 1988); G. Cowlishaw and R. Dunbar, 'Dominance rank and mating success in male primates', *Animal Behaviour*, xli (1991), 1045–56; and L. Ellis, 'Dominance and reproductive success: a review', *Ethology and Sociobiology*, xvi (1995), 253–333. A number of anthropological studies have linked dominance to reproduction; see review in L. Betzig, 'People are animals', in L. Betzig (ed.) *Human Nature: a critical reader* (Oxford, 1997), 1–17. On despotism and democracy in humans and other organisms, see L. Betzig, 'The point of politics', *Analyze & Kritik*, xvi (1994), 20–37; L. Betzig, 'Monarchy', in *Encyclopedia of Cultural Anthropology* (edited by D. Levinson and M. Ember, New Haven, CT, 1996), iii, 803–5; and L. Betzig, 'Why a despot?' in L. Betzig (ed.) *Human Nature: a critical reader* (Oxford, 1997), 399–401.

4 Warfare and population structure

James H. Mielke

Introduction

Wars, epidemics and famines were the three most important precipitating factors in population crises in the past. During these crises, demographic parameters often deviated greatly from their usual levels, changing the structure and composition of human populations. Gutmann (1980) suggests that severe population crises often resulted from several factors working in concert, e.g., war and epidemics. Understanding the dynamics of these types of historical crises, especially warfare, and unravelling their short-term and long-term genetic and demographic effects is a challenging endeavour. The purpose of this chapter is to describe how a war influenced the demographic and genetic structure of the Åland archipelago (Finland) and to show that the impact was shaped and influenced by the population structure and sociocultural features of the area.

The Åland Island archipelago, in between Finland and Sweden, is an excellent location to examine the demographic and potential genetic consequences of a war for a number of reasons. First, the historical data are complete, detailed and accurate, allowing for various types of analyses and studies. Second, the islands are relatively isolated and have been so for years. Thus, it is easy to define the affected population and trace changes in the composition and structure over time and space. Third, the historical records permit a study of the effects that a large military force had on the population, especially as it related to the spread of diseases from the troops to the civilian population. Fourth, the contemporary and historical genetic structure of the islands has been described in detail (e.g. Workman and Jorde 1980; Jorde *et al.* 1982; O'Brien *et al.* 1988a, 1988b), allowing one to suggest potential genetic consequences of the war induced crisis.

Materials

The completeness and diversity of archival sources for the Åland Islands, Finland, permit detailed analyses of historic population changes and dynamics (Pitkänen 1979, 1980; Pitkänen and Nieminen 1984; Pitkänen and Mielke 1993). These extensive archival records were the result of a strong, centralised Swedish government combined with the Lutheran State Church that was attempting to maintain control over the people. Thus, since the seventeenth century the Swedish Lutheran ministers have kept records of baptisms, marriages and burials as well as a general register of their parishioners. Swedish ecclesiastical law of 1686 provided the first *uniform* instructions for record keeping. By the mid-eighteenth century Sweden introduced an extensive system of population statistics (census, age/sex structure). These data complement the vital-events records and parish registers.

The main source in examining the War of Finland (1808–9) is the burial (death) records of the 15 Lutheran parishes that comprise the Åland archipelago. These records provide the date of death and burial, name of individual, his/her occupation, place of residence at death, and the cause of death. To supplement and check our data, we have also used birth (baptism) and marriage records that are similarly detailed. In 1749, parish ministers were required to use these vital-events records to complete yearly statistical tables (Population Change Tables) of births, marriages and deaths. Every five years the ministers had to complete another statistical table on population composition and size (Population Tables). These Population Tables were based on the population registers or Parish Main Books (Communion Books) (Pitkänen and Nieminen 1984).

By the early nineteenth century, the parish registers and population statistics for Åland are considered to be relatively accurate and complete (Pitkänen 1977). There is some under-registration of infant deaths, but these can often be found in the birth records. During the War of Finland, the registration system was not greatly compromised or changed. Several hundred Ålanders temporarily moved to Sweden to escape the Russian occupation. Some of these individuals may have died and were not recorded in the Åland parish books, but most moved back after the war in late 1809. Any omissions would be slight and would not greatly affect the conclusions or interpretations of the data.

The War of Finland

Before examining the demographic impact of the war, a few words about the war itself are appropriate. Tsar Alexander I and Napoleon met in July 1807 to negotiate a peace treaty. One of Alexander's roles in these negotiations was to convince the King of Sweden to participate in a continental blockade against the British. The Swedish government was reluctant to act, and as a consequence, Alexander ordered the invasion of Finland (part of the Kingdom of Sweden at that time). So, on 21 February 1808, Russian troops started their offensive. Southern Finland quickly fell into the hands of the Russians (Osmonsalo 1947). Because of its location in the Gulf of Bothnia, the Åland archipelago provided an ideal staging area for a direct attack on both Sweden and Finland. The Russians first occupied Åland in mid-April 1808, with a small contingency of about 700 men. With the help of a few small warships and their crews, Russian troops were captured by Åland peasants in early May (Schulman 1909). Sweden, realising the strategic location of Åland, started sending troops to the archipelago in late May 1808. Gustav IV Adolf, King of Sweden, then sent a large military force to the islands. These soldiers remained in Åland until being driven out by the Russians in mid-March 1809. Fearing the Baltic ice would melt and leave them stranded, most of the Russian troops returned to mainland Finland at the end of March 1809.

Impact of the war

Demographic changes

After the Great Northern War (1700–21), the Åland Islands contained an estimated population of 6,000 individuals. Åland then grew at an average rate of 0.9 per cent from 1721 to 1800 and 0.5 per cent during the nineteenth century. Just prior to the War of Finland in 1808–9, the population of the archipelago numbered about 13,500 individuals. Growth rates had fluctuated between 0.0 and 1.0 per cent except for a few years (1781–90 and 1806–10) when there were losses (Figure 4.1). The largest loss occurred during the war period (1806–10) when the growth rate was −2.37 per cent. By the middle of 1810 the population consisted of about 11,750 people. These figures suggest that the population was reduced by about 1,750 individuals or 13 per cent. Thus, the war had a major effect on the population in a very short period of time.

Considering that these population losses and changes were associated with a war, one may expect to see major alterations in the population composition and structure as revealed by population pyramids (Figure 4.2). This, in fact, is

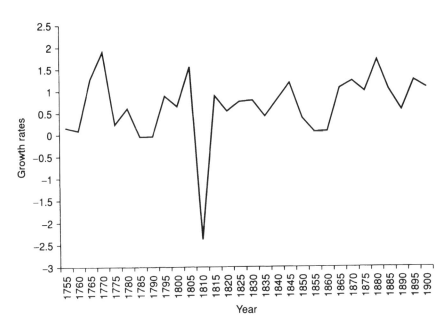

Figure 4.1 Growth rates (%) in Åland, Finland, 1755–1900

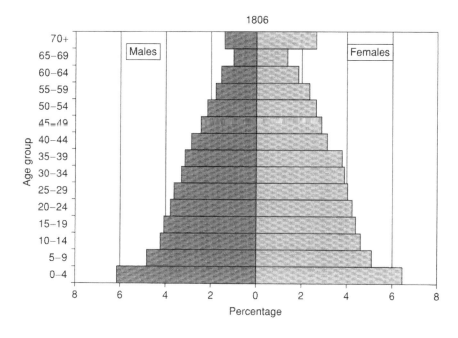

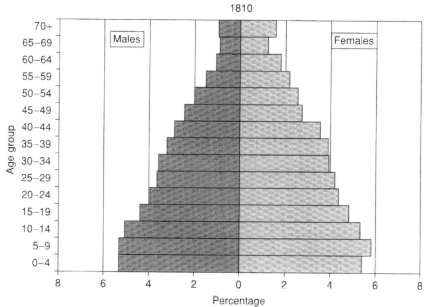

Figure 4.2 Population pyramids for Åland before the 1808–9 war (1806) and after the war (1810)

not generally the case. The population pyramid prior to the war (1806) looks similar to that following the war (1810). A few alterations are apparent. For example, the number of young individuals (ages 0 to 5) was reduced from 12.6 per cent of the population in 1806 to 10.7 per cent in 1810. This change was the result of the high infant and child mortality that occurred during the crisis. There is also a difference in the 70+ age category. In 1806, 4.15 per cent of the population was over 70 years of age, while in 1810 that was reduced to only 2.5 per cent. There are also slightly fewer males aged 50 to 70 after the war than prior, suggesting that mortality was a little greater for older males than for older females. There are no major sex differences as may be predicted during war, famine or other crises periods when there may have been differential exposure to varying risk factors.

The presence of military forces in the Åland Islands had grave consequences for the archipelago's inhabitants in a very short period of time. The crude death rate rose from an average of 29.3/1,000 population (1750–1805) to 49.5/1,000 in 1808 and to 115.2/1,000 in 1809. The crude birth rate dropped from an average of 37.8/1,000 to 31.1/1,000 and to 21.0/1,000 in 1808 and 1809, respectively, influencing the rate of population recovery. The crude marriage rate dropped from 8.2/1,000 to 6.1/1,000 in 1809 and then rebounded to 18.8/1,000 and 11.8/1,000 in 1810 and 1811, respectively. The increase in the marriage rate lasted only two years, settling back to an average rate slightly higher than the period before the war (8.8/1,000 from 1812 to 1831).

When we think of warfare, we often think of the number of causalities sustained on the battlefield. In this historical case, only two deaths can be directly attributed to the fighting that occurred between Russian soldiers and Ålanders, and these deaths occurred in May 1808, before the arrival of the Swedish troops in the archipelago. Most of the deaths during the occupation were the result of diseases that were brought to the archipelago by the Swedish troops who were sent to 'protect' the Ålanders. The monthly number of deaths (Figure 4.3) provides a clear picture of the impact that the Swedish troops had on the civilian population of the Åland Islands during 1808 and 1809. Dysentery makes its appearance in the death records in September and October when there is a slight rise in the level of mortality. The major demographic impact occurred between December 1808 and May 1809 (a six-month period). The losses during this period are attributable primarily to dysentery, typhoid and typhus. Short-term population losses were considerable since the monthly number of deaths was ten times higher than normal. These losses suggest that even short-term, intense mortality crises can have considerable impact on the demographic and genetic structure of a local area.

Historically, large troop concentrations have usually led to economic pressures being placed on the civilian population. Crowding and unfavourable hygienic conditions served as potential sources for the spread of communicable diseases (Gutmann 1977, 1980; Flinn 1980). These types of conditions prevailed during the War of Finland, creating an extremely favourable situation for spreading diseases from the military to the Åland civilians:

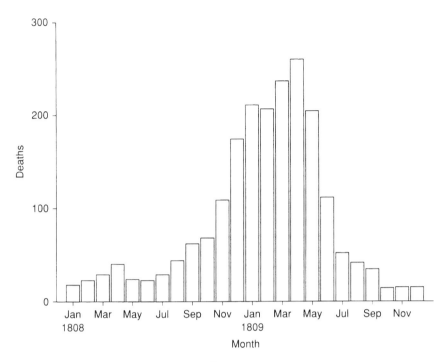

Figure 4.3 Monthly number of deaths in Åland during 1808 and 1809

- By October 1808 the size of the Swedish military contingency numbered about 6000–8000 men. Thus, the number of troops was very large relative to the pre-crisis population of the archipelago, which numbered about 13,500. In addition, the troops were stationed on the Main Island (Fasta Åland), more than doubling the population in some Main Island parishes.
- The Ålanders were in close contact with the troops, thus facilitating the spread of diseases. The authorities required the Ålanders to do numerous things in support of the troops. For example, the civilian population was given the task of transporting sick soldiers to hospitals.
- Starting in October 1808, the soldiers were quartered in the houses of the peasants for the duration of the winter. In many instances 20 to 24 soldiers occupied a single farmhouse with an Åland family. Only a proportion of the sick soldiers were transferred to military hospitals. These actions led to deteriorating hygienic conditions which exacerbated the spread of diseases.
- Military units were periodically transferred from area to area. Thus, new segments of the civilian population were exposed to infectious diseases.
- Three of the eastern parishes (Kumlinge, Brändö and Sottunga) were evacuated between November 1808 and January 1809. These Ålanders were relocated to other parts of the archipelago, increasing the local population density (Anderson 1945).

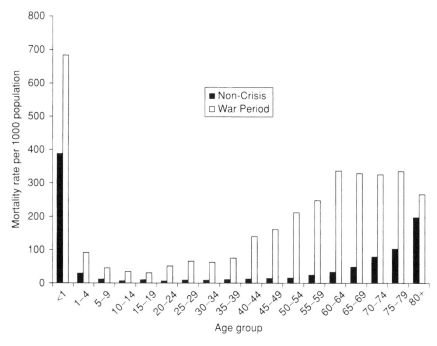

Figure 4.4 Comparison of mortality rates during the war with rates during a 'standard period' (1791–1805)

Because of the excellent archival data for Åland, we were able to calculate age-specific mortality rates for the war period (July 1808 to June 1809). Figure 4.4 graphically depicts the mortality rates for the war period compared to a 'normal' period (1791–1805, a 15-year period just before the war) in Åland. As can be seen, the war period mortality follows a similar pattern as that seen during 'normal' times. The rates are, however, at least twice as high as normal for all age categories (Mielke and Pitkänen 1989). Death rates for infants and young children were especially high. Rates then declined for older children and young adults (up to about age 40 years). The mortality rates then increase rather sharply for individuals over 40 years of age. In fact, the mortality rate in several age groupings exceeds 300/1,000 population, which corresponds to a ten to thirteen-fold increase (Mielke and Pitkänen 1989). This loss of older individuals, especially those between 40 and 60, may have long-term consequences in terms of reproduction and care of young. In general, this pattern suggests that if a population crisis were more intense than the one experienced by the Ålanders, we may expect to see even greater proportional increases in deaths in the 5–64 age groups. This increase would then have major consequences for population growth and recovery. There are also no regular sex differences in age-specific death rates, suggesting that the diseases causing the deaths did not favour one sex over the other or there was differential exposure to pathogens.

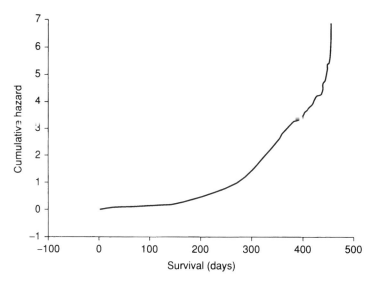

Figure 4.5 Graph of the hazard function during the war of 1808–9

Survival analysis

Cox's Proportional Hazards Survival Models allow one to predict survival times from one or more independent or predictor variables (covariates). Using the covariates of sex, parish and age produced a hazard function that shows a slow initial increase in the hazard of dying during the war period (Figure 4.5). The curve slowly rises for the first 200 days (i.e. until December 1808) after the arrival of the Swedish troops, and then the curve increases rather dramatically. The survival function paints a similar picture of decline in survival and then a trailing off during the last months of the crisis. Median and mean survival times for those who died were just over seven months.

Since the Swedish troops were stationed primarily in the Main Island, we were interested in determining if parish location would show different mortality patterns. The proportional hazards analysis using parish location as a covariate demonstrates that the impact of the crises was disproportionately felt through-out the archipelago, and that one survived longer in different parts of Åland (Figure 4.6). Survival on the Main Island was significantly shorter than in the parishes surrounding the Main Island, which was, in turn, less than in the outer islands. These differences were probably due to differential exposure to the troops and the pathogens they carried. The diversity of impact also appears to be related to ease of movement throughout the archipelago, the relative isola-tion of the outer islands, and historical relationships among parishes. Thus, the regional differences in mortality during the war period may have contributed to the population structure and contemporary genetic heterogeneity seen in the archipelago today.

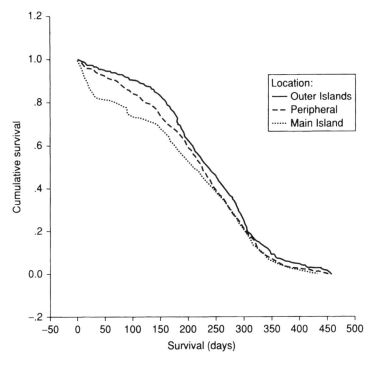

Figure 4.6 Graph of the survival functions for the outer islands (top line), peripheral islands (middle line) and Main Island or Fasta Åland (bottom line) during the 1808–9 war

Discussion and summary

During the War of Finland, mortality on the Åland Islands was intense. The number of excess deaths was about 1,350. These deaths, coupled with a deficiency of about 340 births, caused a short-term loss of about 1,700 Ålanders, or about 13 per cent of the total population. Once Åland's population was reduced by 13 per cent, there is no indication that the effect of the crisis was quickly overcome. In fact, it took about 20 years for Åland to regain its pre-war census (Mielke and Pitkänen 1989). These losses and the heterogeneity of the impact suggests that the war may have been influential in shaping the genetic structure of the Åland Islands.

The genetic structure of a population is a reflection of the contemporary patterns of gene flow and genetic drift as well as past population relationships and historical events. To what extent do demographic crises and other population changes erase or alter the genetic patterns and structure established previously? A number of historical epidemiological studies have shown the effects that infectious diseases have had on the demographic structure of populations. For example, Mercer (1985) has examined the demographic effects of smallpox vaccination in

European populations. Jorde *et al.* (1989, 1990), Pitkänen *et al.* (1989) and Mielke *et al.* (1984) have shed some light on the demographic impact that recurrent epidemics have had on the population and genetic structure of the Åland Islands and mainland Finland. Famine induced crisis mortality that has an impact on the demographic structure of a population may, in turn, influence and shape the genetic structure. Bittles *et al.* (1986) and Bittles (1988) have argued that the Irish Famine of 1846–51 was responsible for creating marked differences in population structure among subdivisions and regions in Ireland. Effects on household size, population density, male/female ratios and effective population sizes were not consistent from region to region despite massive population losses. The genetic consequences of these demographic changes may have increased the rate of random genetic drift. Thus, these researchers suggest that high rates of recessive genetic disorders in the contemporary population can be partially attributed to these famine-induced changes. Smith *et al.* (1990) and Bittles and Smith (1994) have documented the potential genetic consequences of the Irish Famine on population subdivisions in north-eastern Ireland. They found that there were discrete breeding groups with minimal genetic exchange. Coupled with political-religious and geographic barriers to gene flow, this post-famine structure enhanced the potential for genetic drift and inbreeding. Others (Relethford and Crawford 1995; Relethford *et al.* 1997; North *et al.* 1999) argue that recent changes in Ireland, such as the Great Famine, minimally affected the population structure at the county or regional level, but were more likely to impact on the structure of local populations. It is also possible that stochastic local differentiation may conceal spatial patterning at other levels. They realise that these findings neither negate the importance of rapid demographic changes on the genetic structure of human populations nor specifically negate the findings of Bittles and colleagues (Bittles *et al.* 1986; Bittles 1988; Smith *et al.* 1990; Bittles and Smith 1994). Clearly, a long-term and multilevel perspective is essential for understanding the present-day distribution of genes in any area.

In the past historical demographers have attempted to explain mortality patterns and change in terms of living standards and wages (Landers 1993a). These attempts have often been unsuccessful, and now there appears to be some shift in emphasis and focus. A number of historical demographers and epidemiologists now argue that local and regional diversity in the spatial structure (such as population size and density, migration patterns, economic variation) of populations affects the degree and intensity of exposure to infectious agents (Dobson 1992; Landers 1993b; Langford and Storey 1992). These researchers are now focusing their studies at the local and regional level more than the national level. Dobson (1992) even suggests the term 'epidemiological landscapes' to characterise the complex diversity seen in epidemic outbreaks. It is hoped that these types of demographic studies can be combined with temporal and multilevel genetic studies to add new insight into historical population dynamics.

The 1808–9 war-induced population crisis was possibly responsible for some of the genetic diversity that is seen in the Åland Islands today. The exact effects are difficult to determine, but the diversity of the impact of the crisis as shown in the hazard analysis suggests that there should be some genetic heterogeneity among the Main Island parishes and the outer, more isolated island parishes. Genetic studies in Åland have demonstrated greater potential for genetic drift and larger genetic distances among the outer islands than among the Main Island parishes. There are also relatively high levels of genetic differentiation between the Main Island grouping (Fasta Åland) and the outer island parishes (Workman and Jorde 1980; Mielke *et al.* 1982; O'Brien *et al.* 1988a, 1988b; Jorde *et al.* 1982). Selected alleles (e.g. ABO, MN, cde and Duffy) also display inter-regional heterogeneity within Åland, with major differences seen between the Main Island and the outer island parishes (Eriksson *et al.* 1980). Rare alleles (von Willebrand's disease, Hailey-Hailey disease) also appear to be differentially distributed across the archipelago, reflecting initial founding, gene flow and genetic drift (Eriksson *et al.* 1980; Forsius *et al.* 1980; Lehmann *et al.* 1980). All of the local level differentiation seen today is probably the result of historical events (such as founder effect, demographic changes, migration patterns), past relationships among parishes, and the contemporary patterns of gene flow and genetic drift. The 1808–9 war mortality undoubtedly contributed to the contemporary genetic structure of the archipelago by enhancing the potential for drift while at the same time influencing patterns of gene flow (Relethford and Mielke 1994; Mielke *et al.* 1994). The loss of potential mates would possibly increase exogamy rates and spread rare genes further throughout the sparsely populated outer island parishes. At the same time, the higher exogamy rates would lower the potential for genetic drift. For example, the small parish of Sottunga had greater exogamy rates (Mielke *et al.* 1994) and low genetic distances to other parishes, as reflected in allele frequency variation (Jorde *et al.* 1982). Clearly, the specific influences and exact amount of impact that the war had on the genetic microdifferentiation throughout the archipelago would be very difficult to gauge. However, the responses, changes and diversity seen in the Åland archipelago add to the conclusions of Relethford and colleagues (Relethford and Crawford 1995; Relethford *et al.* 1997; North *et al.* 1999) who argue that a multilevel analysis may reveal different structural differentiation and paint a much more complete and comprehensive picture of the genetic structure of a region.

With much of genetics now focused at the molecular level, it would be interesting to speculate about changes in the genetic structure of Åland that could be examine using these new markers and techniques. It appears that mtDNA (mitochondrial DNA) sequence markers and RFLPs (Restriction Fragment Length Polymorphisms) are relatively resistant to demographic shifts in population size, either expansions or reductions (Rogers 1995). However, prolonged and severe bottlenecks and substantial gene flow can alter the overall distribution of mitochondria (Marjoram and Donnelly 1994). VNTRs (Variable Number of Tandem Repeats) and STRs (Short Tandem Repeats) will probably be much more easily affected by gene flow and changes in population size because of the high mutation rates and high variability of these loci. High mutation rate can, apparently, wipe out the past history of changes. The large number of alleles with VNTRs and STRs make them highly susceptible to changes in population size and gene flow (McComb 1996). RAPDs (Randomly Amplified Polymorphic DNAs) appear to be more robust than VNTRs when there are population expansions or reductions. RAPDs are anonymous regions of DNA which are amplified using PCR (Polymerase Chain Reaction) and a single primer. These DNA markers behave in different ways depending upon the extent and magnitude of the demographic changes. Systems such as VNTRs and STRs will rapidly either lose alleles (in population reduction) or gain alleles (during population expansions). Systems such as RAPDs and mtDNA will, on the other hand, change much more slowly and are less affected by small changes in the population size. Severe bottlenecks would, however, wipe out past history (Kidd *et al.* 1991) making them less useful for population structure studies. All of these markers are influenced by gene flow. In the future, it would be very interesting to examine the genetic structure of the Åland Islands using a combination of these systems to gain an additional view of the population dynamics that has shaped the contemporary genetic structure of the region.

Acknowledgements

This chapter could not have been completed without the generous support of many people. First, my thanks go to Dr Malcolm Smith for organising and inviting me to the symposium on *Human Biology and History*. Thanks also to Bjarne Henriksson and his staff at the Landskarpsarkivet, Åland, for their help and understanding. I am indebted to Åland's Lutheran ministers, who allowed us to examine and copy selected archival resources. Thanks are extended to Dr Kari Pitkänen for his comments and suggestions. I also benefited greatly from discussions with Dr Joe McComb about molecular markers and population structure. This research was supported, in part, by a grant from the Sigrid Jusèlius Foundation, Helsinki, Finland.

References

Anderson, S. (1945) 'Öståländska Skärgården Under Kriget 1808–1809', *Budkavlen*, 24, 1–45.

Bittles, A. H. (1988) 'Famine and man: demographic and genetic effects of the Irish famine, 1846–1851', in Bouchet, L. (ed.) *Antropologie et historie ou anthropologie historique?*, Actes des Troisiémes Journées Anthropologiques de Valbonne, Notes et monographies techniques no. 24, Paris: Centre National de la Recherche Scientifique, pp. 159–75.

Bittles, A. H. and Smith, M. T. (1994) 'Religious differentials in postfamine marriage patterns, Northern Ireland, 1840–1915. I. Demographic and isonymy analysis', *Human Biology*, 66, 59–76.

Bittles, A. H., McHugh, J. J. and Markov, E. (1986) 'The Irish famine and its sequel: population structure changes in the Ards Peninsula, Co. Down, 1841–1911', *Annals of Human Biology*, 13, 473–87.

Dobson, M. J. (1992) 'Contours of death: disease, mortality and the environment in early modern England', *Health Transition Review*, supplement to volume 2, pp. 77–96.

Eriksson, A. W., Fellman, J. O. and Forsius, H. R. (1980) 'Some genetic and clinical aspects of the Åland Islanders', in Eriksson, A. W., Forsius, H. R., Nevanlinna, H. R., Workman,

P. L. and Norio, R. K. (eds) *Population Structure and Genetic Disorders*, London: Academic Press, pp. 509–36.

Flinn, M. W. (1980) 'Demography, epidemics, ecology', in fifteenth Congrès International des Sciences Historiques, Bucharest, 10–17 August 1980, *Rapports I, Grands Thèmes et Méthodologie*, Editura Academiei Republicii Socialiste România, pp. 587–616.

Forsius, H. R., Eriksson, A. W. and Damsten, D. (1980) 'Recessive tapetoretinal degeneration with varying diagnoses in Åland', in Eriksson, A. W., Forsius, H. R., Nevanlinna, H. R., Workman, P. L. and Norio, R. K. (eds) *Population Structure and Genetic Disorders*, London: Academic Press, pp. 553–8.

Gutmann, M. P. (1977) 'Putting crises in perspective: the impact of war on civilian populations in the seventeenth century', *Annales de Démographie Historique*, 101–28.

Gutmann, M. P. (1980) *War and Rural Life in the Early Modern Low Countries*, Princeton, NJ: Princeton University Press.

Jorde, L. B., Workman, P. L. and Eriksson, A. W. (1982) 'Genetic microevolution in the Åland Islands, Finland', in Crawford, M. H. and Mielke, J. H. (eds) *Current Developments in Anthropological Genetics, vol. 2, Genetics: ecology and population structure*, New York: Plenum Press, pp. 333–65.

Jorde, L. B., Pitkänen, K. J. and Mielke, J. H. (1989) 'Predicting smallpox epidemics: a statistical analysis of two Finnish populations', *American Journal of Human Biology*, 1, 621–9.

Jorde, L. B., Pitkänen, K. J., Mielke, J. H., Fellman, J. O. and Eriksson, A. W. (1990) 'Historical epidemiology of smallpox in Kitee, Finland', in Swedlund, A. C. and Armelagos, G. J. (eds) *Disease in Populations in Transition*, New York: Bergin & Gravey, pp. 183–200.

Kidd, J., Black, F., Weiss, K., Balazs, I. and Kidd, K. (1991) 'Studies of three Amerindian populations using nuclear DNA polymorphisms', *Human Biology*, 63, 775–94.

Landers, J. (1993a) 'Introduction', in Landers, J. (ed.) *Historical Epidemiology and the Health Transition*, Canberra: Australian National University. Supplement to volume 2 of *Health Transition Review*, pp. 1–28.

Landers, J. (1993b) 'Historical epidemiology and the structural analysis of mortality', in Landers, J. (ed.) *Historical Epidemiology and the Health Transition*, Canberra: Australian National University. Supplement to volume 2 of *Health Transition Review*, pp. 47–76.

Langford, C. and Storey, P. (1993) 'Sex differentials in mortality early in the twentieth century: Sri Lanka and India compared', *Population and Development Review*, 19, 263–82.

Lehmann, W., Forsius, H. R. and Eriksson, A. W. (1980) 'Von Willebrand-Jürgens Syndrome on Åland', in Eriksson, A. W., Forsius, H. R., Nevanlinna, H. R., Workman, P. L. and Norio, R. K. (eds) *Population Structure and Genetic Disorders*, London: Academic Press, pp. 537–45.

McComb, J. (1996) 'The effect of unique history and prehistoric events on the gene pool of the Altai-Kizhi: a study of five variable number tandem repeat (VNTR) loci', manuscript, University of Kansas.

Marjoram, P. and Donnelly, P. (1994) 'Pairwise comparisons of mitochondrial DNA sequences in subdivided populations and implications for early human evolution', *Genetics*, 136, 673–83.

Mercer, A. J. (1985) 'Smallpox and epidemiological-demographic changes in Europe: the role of vaccination', *Population Studies*, 39, 287–307.

Mielke, J. H. and Pitkänen, K. J. (1989) 'War demography: the impact of the 1808–9 war on the civilian population of Åland, Finland', *European Journal of Population*, 5, 373–98.

Mielke, J. H., Devor, E. J., Kramer, P. L., Workman, P. L. and Eriksson, A. W. (1982) 'Historical population structure of the Åland Islands, Finland', in Crawford, M. H. and Mielke, J. H. (eds) *Current Developments in Anthropological Genetics, vol. 2, Genetics: ecology and population structure*, New York: Plenum Press, pp. 255–332.

Mielke, J. H., Jorde, L. B., Trapp, P. G., Anderton, D. L., Pitkänen, K. J. and Eriksson, A. W. (1984) 'Historical epidemiology of smallpox in Åland, Finland, 1751–1890', *Demography*, 21, 271–95.

Mielke, J. H., Relethford, J. H. and Eriksson, A. W. (1994) 'Temporal trends in migration in the Åland Islands: effects of population size and geographic distance', *Human Biology*, 66, 399–410.

North, K. E., Crawford, M. H. and Relethford, J. H. (1999) 'Spatial variation of anthropometric traits in Ireland', *Human Biology*, 71, 823–45.

O'Brien, E., Jorde, L. B., Rönnlöf, B., Fellman, J. O. and Eriksson, A. W. (1988a) 'Inbreeding and genetic disease in Sottunga, Finland', *American Journal of Physical Anthropology*, 75, 477–86.

O'Brien, E., Jorde, L. B., Rönnlöf, B., Fellman, J. O. and Eriksson, A. W. (1988b) 'Founder effect and genetic disease in Sottunga, Finland', *American Journal of Physical Anthropology*, 77, 335–46.

Osmonsalo, E. K. (1947) *Suomen Valloitus 1808*, Helsinki: Söderström.

Pitkänen, K. (1977) 'The reliability of the registration of births and deaths in Finland in the eighteenth and nineteenth centuries: some examples', *Scandinavian Economic History Review*, **25**, 138–59.

Pitkänen, K. (1979) 'Finlands folkmängd är 1749', *Historisk Tidskrift för Finland*, 64: 22–40.

Pitkänen, K. (1980) 'Registering people in a changing society – the case of Finland', *Yearbook of Population Research in Finland*, 18, 60–79.

Pitkänen, K. and Mielke, J. H. (1993) 'Age and sex differentials in mortality during two nineteenth century population crises', *European Journal of Population*, 9, 1–32.

Pitkänen, K. and Nieminen, M. (1984) 'The history of population registration and demographic data collection in Finland', in Majava, A. (ed.) *National Population Bibliography of Finland 1945–1978*, Helsinki: Finnish Demographic Society, pp. 11–22.

Pitkänen, K., Mielke, J. H. and Jorde, L. B. (1989) 'Smallpox and its eradication in Finland: implications for disease control', *Population Studies*, 43, 95–111.

Relethford, J. H. and Crawford, M. H. (1995) 'Anthropometric variation and the population history of Ireland', *American Journal of Physical Anthropology*, 96, 25–38.

Relethford, J. H. and Mielke, J. H. (1994) 'Marital exogamy in the Åland Islands, Finland, 1750–1949', *Annals of Human Biology*, 21, 13–21.

Relethford, J. H., Crawford, M. H. and Blangero, J. (1997) 'Genetic drift and gene flow in post-famine Ireland', *Human Biology*, 69, 443–65.

Rogers, A. (1995) 'Genetic evidence for a Pleistocene population expansion', *Evolution*, 49, 608–15.

Schulman, H. (1909) *Taistelu Suomesta 1808–1809*, Porvoo, Suomi-Finland: Werner Söderström.

Smith, M. T., Williams, W. R., McHugh, J. J. and Bittles, A. H. (1990) 'Isonymic analysis of post-Famine relationships in the Ards Peninsula, N.E. Ireland: genetic effects of geographical and politico-religious boundaries', *American Journal of Human Biology*, 2, 245–54.

Workman, P. L. and Jorde, L. B. (1980) 'The genetic structure of the Åland Islands', in Eriksson, A. W., Forsius, H. R., Nevanlinna, H. R., Workman, P. L. and Norio, R. K. (eds) *Population Structure and Genetic Disorders*, London: Academic Press, pp. 487–508.

5   Isonymy analysis

The potential for application of quantitative analysis of surname distributions to problems in historical research

Malcolm Smith

Introduction

The purpose of this chapter is to suggest that some techniques, widely used in biological anthropology to interpret surname distributions in terms of population structure, might find wider application to help answer questions generated by historians. 'Population structure' (Morton 1982) has a particular meaning in the context of the genetics of populations, and refers to the distribution of genes and genotypes, and the correlates of this distribution. The term is used almost synonymously with 'genetic structure' or 'population genetic structure'. Migration, mate choice and the subdivision of the population by geographical or socio-economic factors may all influence population structure, by affecting the stability and turnover of population, migration and mate choice, and the local distribution of 'kinship', which in biological anthropology has an explicitly genetic definition (see, for example, Morton 1982; Morton *et al.* 1971), though it retains much in common with a more intuitive construction of the word. Historians, too, have long been interested in the stability, turnover and structure of local populations, as either a purely descriptive aspect of local population studies, or to explain socio-cultural and economic change and diversity. This chapter sets out to illustrate that isonymy methods – a set of quantitative techniques for the analysis of surname distributions, developed from the theory of population genetics – can be more broadly applied to answer historical questions which are not themselves couched in terms of genetics, but rather generated from within historical debate.

Historians already have an interest in and familiarity with the distribution of surnames, and the chapter begins with a brief overview of approaches to surname studies in history. This is followed by a description of surname methods used in biological anthropology, including the theory and assumptions which underlie them. Case studies will then be used to illustrate one particular isonymy method and the intuitive inferences that can be drawn from the results, and the chapter concludes by exploring the kinds of historical questions which might be tractable to the analyses described here.

Surname studies in history

Surname studies may be divided broadly into those with a primarily philological emphasis (e.g. Reaney 1967) whose focus is the etymology of names, and those which use surnames to generate some wider historical interpretation. Frequently such interest has been in surnames' geographical (and textual) origin and distribution, with especial interest paid to names which seem to be limited in distribution to a few kilometres of their origin, indicative of stability of population. Hoskins (1972) was struck by the implications of this property, drawing on the work of McKinley (1969) and on unpublished results of Redmonds from Yorkshire and Blewett from Cornwall. Hoskins hailed the study of surname distributions as a field of immense promise, and subsequently a great deal has indeed been achieved, notably by Hanks (Hanks 1992–3; Hanks and Hodges 1988), Hey (1993, 1997, 2000) and the authors of the English Surname Series, which has done much to reveal the historical continuity and pattern of names at a county level. The volumes so far published cover the West Riding of Yorkshire (Redmonds 1973), Norfolk and Suffolk (McKinley 1975), Oxfordshire (McKinley 1977), Lancashire (McKinley 1981), Sussex (McKinley 1988), Devon (Postles 1995) and Leicestershire and Rutland (Postles 1998). At a national level Rogers (1995) and Rowlands and Rowlands (1996) have used the chronological and spatial distribution of names to demonstrate migration patterns in historical England and Wales, respectively. Historical geographers might also have been expected to show an interest in the spatial and temporal distribution surnames, but in fact their endeavours in this field have been rather limited, despite the advocacy of some researchers who have used the origin and distribution of individual names to estimate migration (Porteous 1982; Giggs 1994).

Hey (2000) gives an overview of the approaches employed in researching the historical origins and distribution of names, and argues convincingly that a combination of etymological and genealogical evidence seems to be the most reliable method of tracing a surname back to its origin. In particular, Hey's use of genealogical techniques linking documentary sources from different time periods gives greater confidence to assertions of surname derivation and dispersal.

On a national scale, Lasker and Mascie-Taylor (1985) and Boldsen *et al.* (1986) examined the pattern of selected names, and Hanks (1992–3) used telephone directories from the 1980s to derive the distribution of the 15,000 most frequent surnames. The geographical pattern detected by Hanks from this source appears to be as marked as it was a hundred years previously when Guppy pioneered the quantitative analysis of surname geography, recording and commenting upon the frequency of farmers' names taken from Kelly's Post Office Directories (Guppy 1890). The present-day distribution of names, and a comparison of the contemporary pattern with its spread from the putative or ascertained point of origin, often prompts the further observation that dispersal of many names through the six hundred or so years during which they have been inherited is geographically very restricted, with many names still concentrated close to their apparent point of origin. This suggests a stability of population which is at odds with the conventional wisdom of historical demography at a parish level, which envisages high levels of migration and population replacement at a parish level since medieval times. One objection to the surname evidence is that it is selective, that is, the stable surnames are the ones that are the most favoured illustrations. On the other hand, the surname and parish demography evidence may differ because they represent contrasting scales of analysis, in which case the conflicting interpretations need careful attention. This question of the scale of local community and the sphere of social interaction will be taken up again in the final section of the chapter.

While in the research described above the surnames themselves have been the primary focus of interest, a number of studies have attempted to include analyses of local surname sets in the context of a broader study in order to elucidate the stability or transience of local population. For example, Howell (1983), in her study of the Leicestershire estate of Kibworth Harcourt, presented an illustration showing the presence of individual surnames through time, which gave a clear visual impression of the chronology of persistence, loss and introduction of new names. A similar graphic was used by Stapleton (1999) to illustrate persistence of lineages in Odiham, Hampshire in relation to decisions about the inheritance of land between generations. In order to measure the constancy of local populations, Buckatzsch (1951) devised a statistic in which the proportion of names recorded in some time period and persisting from a previous period was tabulated as a triangular matrix. This method introduced a quantitative technique, counting presence or absence of a surname, but not its frequency. This simple but effective approach has been used or reinvented on a number of occasions (e.g. Watson 1975; Pollitzer *et al.* 1988; Wyatt 1989; Halse 2000). Coleman reviewed the evidence from a number of parishes for population turnover from surnames in a wide-ranging paper on various aspects of historical mobility (Coleman 1984).

Working within the theoretical framework of biological anthropology, Lasker (1977) devised a measure of relationship based between two populations based on the frequency of every surname in both populations. This and similar measures will be discussed in greater detail later, in relation both to theory and case studies. I mention it early here, because while it has been widely used in biological anthropology, it has also been employed occasionally by historians. David Souden worked with Lasker on an analysis of the Marriage Duty Returns from Kent (Souden and Lasker 1978). Daniel Scott Smith (1989) used isonymy to analyse tax listings to explore the structure of kinship in a number of towns in colonial Massachusetts and David Postles (1995) has used the statistic in his English Surname Series monograph on Devon. Lasker's measure shares an important property with Buckatzsch's statistic, which is that both are in principle inclusive of all surnames in the local population. This contrasts with many distributional and all genealogically based studies, which focus on the behaviour of individual surnames, and thus present an exemplary and perhaps biased sample of the behaviour of surnames.

Surname studies in biological anthropology

One of the preoccupations of those studying the genetic structure of populations has been the estimation of rates of consanguineous marriage, and the ascertainment of the biological consequences of inbreeding (Bittles 1994, 1998; Bittles and Neel 1994) and it was in pursuit of the study of inbreeding that surname analysis entered biological anthropology. As early as the nineteenth century, researchers were collecting data to ascertain the effects of consanguineous marriage on offspring, with 172 such books and papers indexed by Huth (1879), including one study which made use of surname frequencies for this purpose conducted by G. H. Darwin (1875), son of Charles Darwin. There was also a further notable contribution to the estimation of cousin marriage from surname frequencies by Bramwell (1939), but the foundations of the systematic use of surnames to estimate inbreeding and explore population structure were laid with the publication of the technique and theory of marital isonymy by Crow and Mange (1965).

Marital isonymy

Crow and Mange (1965) showed that the rate of isonymous marriage (that is marriage where the bride and groom have the same surname before marriage) could be used to estimate the mean inbreeding coefficient of the population, and that this estimate of inbreeding could be partitioned into random and non-random components, reflecting demographic constraints on the one hand and behavioural choice on the other. The paper contained a case study of the North American religious isolate, the Hutterites, with a comparison of inbreeding rates derived by conventional methods from pedigrees (Cavalli-Sforza and Bodmer 1971) with those obtained from isonymy.

Crow and Mange's method is based on their observation that for a wide range of types of consanguineous marriage, there is a constant relationship between the probability of such a marriage being isonymous, and the inbreeding coefficient of the offspring. Thus, $F = I/4$, where F is the inbreeding coefficient and I is the expected proportion of isonymous marriages in the particular category of marriage.

One of the common estimators of inbreeding based on isonymy is as follows:
The random expectation of isonymy, or *random isonymy*, is $\sum p_i q_i$ and so random inbreeding is given by

$$F_r = \sum p_i q_i / 4$$

where p_i is the relative frequency of the ith surname among the brides and q_i is the relative frequency of the ith name among the grooms, and the summation sign $\sum$ means that coefficient results from the addition of pq calculated for every surname in turn.

The non-random (F_n) and random (F_r) components of total inbreeding (F) can be estimated using theory formulated by Wright (1951):

$$F = F_n + (1 - F_n)F_r$$

where $F_n = (I - \Sigma p_i q_i)/4(1 - \Sigma p_i q_i)$.

Alternative estimates are given by Crow (1980) and Relethford (1988).

The theoretical model which underlies this interpretation makes the analogy between surnames and genes, with each surname representing a different allele at a single gene locus. If surnames are to stand as quasi-genetic markers in this way then certain assumptions about the distribution and dynamics of names have to be made.

First, surnames should be inherited in a systematic fashion that mimics genetic transmission. Second, in order to interpret like surnames in terms of genetic relationship, names should be monophyletic (all present-day copies of a name descended from a single common ancestor).

Third, for Crow and Mange's marital isonymy, there should be an equal likelihood of occurrence of matrilineal and patrilineal kinship within marriages of any degree of relationship. The reasoning is as follows. In the case of marriages between first cousins, for example, a man may have four different categories of marriage partner: his bride may be his father's brother's daughter (FBD), father's sister's daughter (FZD), mother's brother's daughter (MBD) or mother's sister's daughter (MZD). Thus, the siblings who are parents of the consanguineous couple are either brother-brother, brother-sister, sister-brother or sister-sister. If each of these marriage-types occurs with equal frequency then one in four will be isonymous (the situation when the bride is FBD). There may, however, be a preference for one marriage type over the others (e.g. if marriage between cross-cousins (FZD, MBD) is preferred to marriage between parallel-cousins (FBD, MZD)), or there may be a bias towards one type as a side-effect of some other process. For example, if men remain in their birth area (virilocal settlement) and females disperse through emigration, then FBD will be more frequent and MZD less frequent than expected. Similarly, since the age difference between potential mates is a constraint on marriage and since men have the capacity for a longer reproductive span than women, the potentially greater difference in age between the children of brothers compared to the children of sisters will favour MZD marriages over FBD, biasing the inference of inbreeding.

The first assumption is clearly crucial to the genetic analogy, and isonymy is practicable only where regular pattern of surname inheritance occurs; it is clear that in populations without inherited surnames (Iceland, for example) the exercise would be meaningless. Many European and European-derived systems of naming follow a regular pattern of paternal inheritance, which is analogous to the inheritance of the Y chromosome. Iberian names are inherited both paternally and maternally, and in these populations isonymy can be calculated using both elements of the family name (Pinto-Cisternas *et al.* 1985). Change in individual names over time is analogous to the genetic process of mutation, and in periods where such change was especially rapid (in written records before the spellings were stabilised, or the transformation of European or Jewish names in the USA or UK, for example) kinship will be underestimated.

It is the second assumption above which has given most cause for concern, however, with the recognition that many names – names derived from occupations, for example – have multiple origins. It is a salient point, then, that Hey (2000) confirms an observation repeatedly made by workers using the genealogical and distributional method, namely that a large number of names do appear to derive each from a single origin. Surnames derive generally from placenames and landscape features, occupations, patronymics or nicknames. Of these categories, only placenames were reckoned reliably to give rise to names with a single point of origin, but recent research using the genealogical method (Hey 2000) suggests, at the very least, that some names derived from topography may each be descended from a single ancestor.

Further evidence of the robustness of the second assumption comes from studies of the association of particular surnames with Y chromosome haplotypes. The surname Cohen has been shown to have a very restricted range of Y chromosome diversity, consistent with its inheritance by the Jewish priestly caste (Skorecki *et al.* 1997; Thomas *et al.* 1998). In Ireland, individuals with Irish surnames of presumed Celtic origin were shown to have a distinctively high frequency of certain related Y chromosome haplotypes compared to those with Scots-Irish or English-derived surnames (Hill *et al.* 2000). Even more specifically, Sykes and Irven (2000) showed that a large sample of men bearing the name Sykes (a name from the West Riding of Yorkshire derived from a topographical feature) shared a single Y chromosome, with variation which would be consistent with non-paternity at one per cent per generation. Unpublished research by Sykes commissioned for a series of BBC radio programmes by George Redmonds entitled *Surname, Genes and Genealogy* (Redmonds 2001) showed that the bearers of a number of other names, for example Dyson (Poole 2001) also shared close Y chromosome relationships (Hey 2001). However, Jobling (2001) has commented that heterogeneity of origin cannot be confidently excluded unless a larger number of Y chromosome loci are tested than is currently the case.

Besides these constraining assumptions, Lasker (1985) has pointed out a further drawback to inference from marital isomymy: the number of such marriages is always small, so that sampling error is an impediment to reliable inference.

Following Crow and Mange, the subject has been developed in a number of directions, some exploring alternative theoretical frameworks for surname study (e.g. Kosten and Mitchell 1990) others concentrating on applications and extensions of Crow and Mange's original insight. Useful reviews are provided in a special section of the journal *Human Biology* edited by Gottlieb (1983), in Lasker (1985) and Crow (1980, 1989).

To avoid the problems of sampling error mentioned above, Lasker and Kaplan (1985) devised an alternative statistic for the detection of mating structure within a population, the calculation of Repeated Pairs. This quantifies the extent to which there is a recurrence of marriages between persons with the same pair of *different* surnames. It thus looks for persistent networks of intermarriage rather than inbreeding per se. Values of the statistic based on the underlying composition of the population have been calculated (Chakroborty 1985) and there have been a number of applications of the method (Lasker *et al.* 1986; Mascie-Taylor *et al.* 1987; Pollitzer *et al.* 1988; Koertvelyessy *et al.* 1992) and a review (Relethford 1992).

Random isonymy

In Crow and Mange's original paper, *random isonymy* connotes a generalisation from the incidence of particular marriages between people of the same name, to the potential for marriages between people of the same name. This potential for isonymous marriage is based on the frequency of surnames among the brides and grooms of the same population rather than on actual marriages, assuming that all brides and grooms are potentially available to each other as marriage partners. It was used by Crow and Mange (1965), employing the relationship $F = 1/4$, to estimate the random component of inbreeding.

$$F_r = \Sigma p_i q_i / 4$$

Random isonymy (or a simple transformation of it) has subsequently been championed as a measure in its own right, offering an insight into the 'kinship' (Morton 1973, 1982; Morton *et al.* 1971), 'relationship' (Lasker 1977) or 'genetic distance' (Relethford 1988) between populations. Lasker's coefficient of relationship by isonymy (R_i) is formulated as *relationship* rather than *inbreeding*:

$$R_i = \Sigma p_{iA} \cdot p_{iB} / 2$$

where p_{iA} is the relative frequency of the ith name in population A and p_{iB} is the frequency of the ith name in population B.

The contrast between Crow and Mange's and Lasker's calculation exactly parallels the observation that the *inbreeding coefficient* of the offspring of a consanguineous marriage is half the size of the *coefficient of relationship* between the consanguineous parents. For example, first cousins have one-eighth of their genes in common, and the inbreeding coefficient of their offspring is one-sixteenth.

For Lasker's coefficient of relationship and other measures of random isonymy between populations to be valid the first and second assumptions above must be met. Analogous to the third assumption above is the further assumption that the patterns of male and female migration between the groups must be equivalent, as a hypothetical example will show. Consider two nepotistic occupations, say fishing and farming (relevant because groups of this nature occur in the case studies below). Suppose inheritance customs mean that boats and farms are always passed from father to son, and that the groups differ initially in their surnames and in their gene frequencies. If the groups are endogamous then their name and gene frequencies will remain distinct. If, however, farmers' sons marry fishermen's daughters and fishermen's sons marry farmers' daughters, the genetic differences between the groups will quickly disappear, but the distinction on the basis of surnames will persist. For surnames to reflect genetic structure accurately, then, women must migrate and marry between groups to the same extent as men do.

The conceptual shift from marital to random isonymy has two important consequences. The first is the changing horizon for the scale of analysis. Whereas inbreeding from isonymy necessarily focuses on a breeding population, random isonymy places no such constraint on the surname distributions to be compared. To be sure, subdivisions of a population within a parish can be investigated, but there is no limit to the parish as the locus of study. In a strict sense, we are always considering kinship *within* a subdivided population, but in practice we can make comparison *between* populations. Second, the shift of emphasis from marital to random isonymy frees the method from its close association with marriage, permitting the relationship between populations to be interpreted in terms of broader determinant processes, which have mate choice or migration as outcomes, such as geography, religion, ethnicity and socio-economic factors.

Moving from the narrow focus on marriage also has a practical consequence, which is that random isonymy can be calculated from any lists of names, and does not require that marriage records be the data source. Typically, the method has been applied to data from vital records, censuses and other population listings such as telephone directories and electoral registers, though each data source has to be considered on its merits as regards accuracy and bias. The number of studies using surnames in the above way as genetic markers is burgeoning, with hundreds of publications since the mid-1970s, not least because the data are fairly quick and cheap to collect and analyse. However, the fact that surname data are available from a wide range of populations throughout long historical periods, when systematic sources of other genetic or quasi-genetic information are simply unavailable, means that they have a value beyond mere expedience.

Case studies

The case studies are intended to illustrate the use of surname distributions as an aid to revealing the structure of historical populations. There are three components of structure that can be distinguished, the first dependent on distance, the second on time, and the third on non-spatial constraints, for example social class, occupation and religion. As regards distance effects, there is rapidly accumulating evidence that in surname distributions show a close correspondence to geography. This has been demonstrated both at a local level (over distances of a few kilometres), for example in the nineteenth-century population of the Ards Peninsula, County Down, Northern Ireland (Smith *et al.* 1990), in other parts of the UK (Holloway and Sofaer 1989; Lasker 1997, 1999) and also over much larger distances in Europe (Barrai *et al.* 1996, 1999; Rodriguez-Larralde *et al.* 1998a, 1998b) and the Americas (Barrai *et al.* 2001; Christensen 1999; Rodriguez-Larralde and Barrai 1997; Rodriguez-Larralde *et al.* 1993, 2000).

In the case studies below the emphasis is on the revelation of temporal and non-spatial aspects of population structure. The particular question addressed is this: does the existence of distinctive occupational groups within a local population constrain mobility and marriage between them to the extent that subdivisions of the population will be revealed by the surname distributions of those groups?

The method employed is the same in all the following examples. Lasker's coefficient of relationship by isonymy (R_i) is computed between populations from lists of men's surnames taken from nineteenth-century censuses. The resulting matrix of coefficients is displayed as a two-dimensional 'map' by a computer-based data reduction technique called non-metrical multidimensional scaling (MDS). This is just one of a range of possible techniques available to reduce a triangular matrix of coefficients to two dimensions, and is implemented in Statistical Package for the Social Sciences (SPSS) and other statistical program packages. The reason for the data reduction is that it allows the relationships between populations to be easily visualised, by comparison with the difficult task of interpreting the relationship matrix itself. The simplification has a price, however, as there will be some distortion of the two-dimensional plot of relationships. The distortion is measured either as 'stress', or as R^2, the proportion of variance in the unreduced data explained by the two-dimensional solution.

Case study: coastguards at Selsey

This first example is chosen because it enables the pattern of surname distributions to illustrate something which is clearly suggested from other information. It is intended, then, as a demonstration of the method and to show its efficacy. The area of study is Selsey, on the Sussex coast. The population was chosen because it is one where there were few occupations each undertaken by rather many men, with some of the same occupational groups as in another coastal population, the parish of Fylingdales, North Yorkshire, which features in the second case study.

Additionally there were in Selsey between 25 and 30 coastguards in every census year between 1841 and 1881. The Coastguard had long had close connections with the Royal Navy, both in terms of administration and recruitment and in 1856 the Coastguard came under the direct control of the Admiralty. The men serving as coastguards were in short-term postings and could usually be expected to come from outside the local community. These are a useful group of men for purposes illustrative of surname distribution, since in any local context we should expect them to be an alien and transient component of population.

Selsey is situated close to the English Channel, on the peninsula of the Manhood, 15 km south of Chichester. Built upon a ridge of raised land, it is surrounded by sea on two sides and on a third (to the north) by low marshy ground. In the nineteenth century, Selsey was little more than a small village – the biggest population at census was about 900 – sustained almost entirely by agriculture and fishing. The Selsey peninsula provided a wide variety of fish including 'Selsey cockles' and other shellfish. With the closure of the harbour in 1873, the cockle industry was finished, and though dredging for oysters also declined towards the end of the nineteenth century, the annual income from fish sold locally was still considerable.

The local historian of Selsey, Heron-Allen, found that 'the fishermen of Selsey form a curiously isolated and independent body, having little or nothing to do with the village proper' (Heron-Allen 1911: 323). Though there was no guild of

Table 5.1 Census numbers of working men in Selsey

Occupation	Census year				
	1841	*1851*	*1861*	*1871*	*1881*
Fisherman	70	83	89	80	79
Farmer	16	18	10	16	24
Ag. lab.	85	119	133	121	99
Coastguard	25	27	26	28	28
Other	54	51	64	76	70

fishermen, in practice they operated a closed shop, and a 'foreigner' (i.e. a man from another parish) could join the trade initially only as a mate to an established fisherman and thence, gradually, work his way into the community. We have no equivalent particulars of employment in agriculture in Selsey, although, like the Chichester district as a whole, it was prosperous in the mid-nineteenth century, and the new corn market, built in Chichester in 1835, became an important trading centre. The only other noteworthy employer in Selsey at the time was a Mr Pullinger, whose mousetrap factory was established in the 1850s. Although this enjoyed a considerable reputation, with sales of over 1 million mousetraps by 1882, it employed a maximum of only 40 workers, and of these only a minority are to be found in the Selsey censuses – 15 in 1861 and 17 in 1871 (Wolf 1882).

The surname and occupation of each working man was extracted from microfilm copies of the census enumerators' books for 1841–81. Females were excluded from the analysis because of the peculiar distribution of women's occupations and the fact that maiden names are not recorded in the census. The men were grouped into the following occupational categories: agricultural labourer, farmer, fisherman, coastguard and the catch-all category 'other' comprising mainly tradesmen, craftsmen and professionals. 'Agricultural labourer' included the related jobs 'farm labourer', 'farm servant', 'shepherd', 'hind', etc.

The data set is not large, fewer than 1,500 cases all told, and Table 5.1 shows a breakdown by decade and occupation. Clearly some chance variation might be expected among the farmers and coastguards, as their numbers are few. For the data amalgamated over all years Lasker's coefficient of relationship by isonymy (R_i) was computed between each occupational group. An MDS plot of these relationships is shown in Figure 5.1. The point to emphasise is that the coastguards, the one group we expect to be derived from outside the local community, are shown in the plot to be the most remote in terms of their surname distribution. The next analysis computes R_i between each job category in every decade, and the MDS plot of these relationships is shown in Figure 5.2. The data points for each occupation in successive decades have been joined by a line with a directional arrow indicating the direction of time. Again, let us concentrate on a simple inference about the coastguards compared to the rest; their short-term postings and replacement by new men, is reflected in the way their surnames

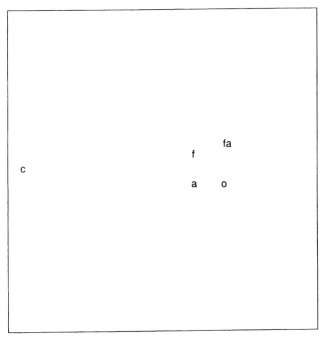

Key
a   agricultural labourers
c   coastguards
f   fishermen
fa farmers
o   others

Figure 5.1 MDS plot of Lasker's R_i, Selsey, all years combined

differ greatly in Figure 5.2 from decade to decade. The points are wider spread, and consequently the total length of the arrow is much longer than for the other occupations. That this is not due to sampling effect is suggested by comparison with the farmers, a smaller group of people than the coastguards, and thus more susceptible to sampling fluctuation, but whose surname distributions remain much more constant throughout the period. The point of this example is to show how a particular and very clear historical situation is illustrated through Lasker's method and MDS by a simple summary diagram.

Case study: occupations in Fylingdales

In the next case study we shall attempt to draw inferences from a historical situation of greater complexity.

The parish of Fylingdales lies on the north Yorkshire coast, just south of Whitby. In the second half of the nineteenth century the economic focus of Fylingdales was the thriving little town of Robin Hood's Bay, known locally as 'Bay Town'

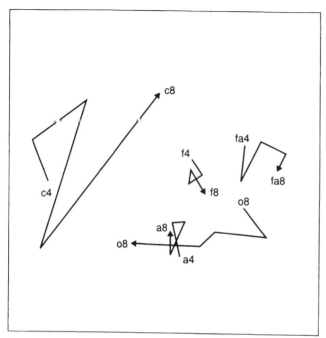

Key

a4–a8   agricultural labourers
c4–c8   coastguards
f4–f8   fishermen
fa4–fa8 farmers
o4–o8   others

Figure 5.2 MDS plot of Lasker's R_i, Selsey, 1841–81

or just 'Bay'. Sailors and fishermen were numerous among the male population, which otherwise mainly comprised tradesmen, craftsmen and professionals.

The town's early history is obscure, but it seems from the first to have been a fishing village. There is no mention of it in the Domesday Book, but a discussion of tithe duties, probably in 1245, says that the Bishop of York received a tithe of fish taken at Robin Hood's Bay and Whitby (Young 1817). Leland (1964) visited this coast in 1538 and recorded his journey:

> Thens [from Scarborough] to a fisher tounlet of 20 Bootes caullid Robyn Huddes Bay, a Dok or Bosom of a Mile yn lenghth; and thens 4 Miles to Whitby . . . a great Fischar Tounie.

By 1800 Robin Hood's Bay, along with Staithes and Runswick, far exceeded Whitby as a fishing port; Young (1817) observes that Whitby had but nine fishermen, and that its daily fish market was supplied by those three villages. By the mid-nineteenth century seafaring employed more men than fishing – 136 sailors to 38 fishermen at the 1851 census. Merchant seamen and master mariners from Bay Town had long been serving in the Whitby-based coastal and international shipping enterprise. Half the thousand ships that sailed in and out of Whitby each year were registered there, and part-ownership of vessels, sold in shares of one-sixty-fourth, provided an opportunity for modest but rewarding investment for the more prosperous. A number of such people lived in Bay Town, and are designated as 'shipowners' in the census enumerators' books.

Not all the parish looked to the sea for its livelihood; the rural hinterland supported a substantial agricultural community centred on the hamlets of Thorpe (Fylingthorpe) and Raw, and also dispersed among many isolated farms and cottages. Additionally the alum works quarries at Stoupe Brow and at Peak employed labourers from 1615 until the closure of the Peak alum works in 1858. At the 1851 census the coastal and inland populations were approximately equal, with 871 people in Robin Hood's Bay and 894 inland, comprising 215 in Thorpe, 143 in Raw, 189 at Stoupe Brow and 347 in the scattered rural dwellings.

Some indication of the social isolation even within this small parish is provided by the descriptions of early-twentieth-century life given in Leo Walmsley's novels, in which Bay Town, Walmsley's childhood home, goes by the name of 'Bramblewick'. Bramblewick folk called incomers (like Walmsley himself) 'Foreigners', and boys from Bramblewick could not meet lads from the adjacent hamlet Thorpe without free flow of insults, blows and blood.

> There was a standing quarrel between the boys of Bramblewick and the boys of Thorpe. Nearly all the boys of Bramblewick were the sons of either fishermen or sailors. Those of Thorpe were mostly farmers' sons. They didn't wear jerseys, but corduroys which stank, and they wore clogs in winter. We called them Thorpe cloggers and they called us Bay bumpers. If ever a boy from Bramblewick went to Thorpe and a clogger saw him there'd be a fight straight away, and the same thing happened if a clogger ventured into our village, unless his father happened to be with him, in which case we'd just call after him 'Thorpe clogger, Thorpe clogger'.
>
> (Walmsley 1997: 13–14)

This is an interesting passage because it links together several aspects of life in the parish, all of which might have some influence on the isolation or integration of components of the community – geography, occupation and even material culture, the different clothes worn by the different occupational groups. This last subject can be further illuminated by looking at the celebrated photographs taken by F. M. Sutcliffe, which clearly depict the differences alluded to by Walmsley.

In a study of the 1851 census, the parish population was shown to be considerably differentiated in terms of surname distributions, and it was argued that there were both spatial and occupational components to this variation (Smith and Hudson 1984). In order to see whether the occupational differentiation persisted over time, the analysis was extended to include all the censuses 1841–81. The data were taken from published transcripts of the 1841, 1851 and 1861 census

Table 5.2 Census numbers of working men in Fylingdales

Occupation	Census year				
	1841	1851	1861	1871	1881
Mariner	91	136	130	111	96
Fisherman	39	39	27	17	18
Shipowner	11	13	22	26	13
Farmer	100	68	99	90	110
Ag. lab.	108	79	126	95	65
Other	97	134	138	150	142

enumerators' books and from unpublished transcripts of the 1871 and 1881 censuses. As in Selsey, the evidence is confined to resident men with occupations, or to the evidence of such men in their absence (e.g. 'mariner's wife', 'farmer's widow'). The designation 'mariner's wife' is first seen in the 1851 census. In 1841, the fact that men were away at sea was inferred from the large number of single women with families, a phenomenon for which it is not, of course, the only explanation. In this case, cross-referencing between censuses is the most reliable method of ascertainment, but it produces a bias towards recognition of persistent families. More difficult is the identification of absent mariners whose wives do not have children living at home, and at all periods there is the problem of overlooking single mariners away at sea.

For each census year, men were assigned to one of the following occupations: farmer, fisherman, mariner, shipowner, agricultural labourer and 'other'. The numbers of each are shown in Table 5.2. Lasker's R_i was calculated between groups within and between census years, and the matrix of pairwise relationships reduced to two dimensions by MDS. These relationships are shown in Figure 5.3, which has been drawn so that the points for each occupation in successive decades are joined by an arrow through time.

The most noteworthy general feature of this plot is the clear division of the field into three clusters: a 'maritime' group, including fishermen, mariners and shipowners; an 'agricultural' group, including farmers and agricultural labourers; and 'others'. These clusters show that the occupational groupings remained distinct throughout the 50-year period. The population structure inferred from surname distributions appears to persist, though with some modification of detail which allows further interpretation.

The tightest cluster, indicative of the most stable population, is that of the maritime trades. Within these, the mariners show a regular and gradual shift through time towards the centre of the plot. The shipowners, perhaps because they are the smallest group of people (maximum 26 in 1871) and thus subject to sampling variation, show no such clear pattern, but neither do they vary very much. The fishermen, too, are a small group, but remarkably free from sampling variation, with the same points almost superimposed in successive decades. This occupation tends to be hereditary, and from the evidence of the censuses it

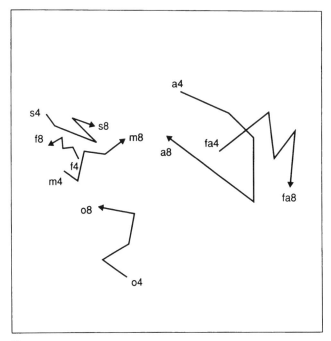

Key
a4–a8   agricultural labourers
f4–f8   fishermen
fa4–fa8   farmers
m4–m8   mariners
s4–s8   shipowners
o4–o8   others

Figure 5.3 MDS plot of Lasker's R_i, Fylingdales, 1841–81

would appear that in Fylingdales, as in Selsey, it is not an occupation that readily admits newcomers. More than any other group, the fishermen are often represented by the same individuals from decade to decade, and the censuses shows their mean age to increase from 43.5 to 62.6 during the study period. This is the profile of a way of life that was in decline, with the old men carrying on, and very few young men entering the trade.

The agricultural cluster is looser, and although there is much general overlap between the agricultural labourers and farmers their patterns differ in some important details. The farmers show a trend in time towards the edge of the plot, suggesting that the farms were falling into the hands of landowners with surnames new to the parish. The agricultural labourers vary much more from decade to decade than the farmers or any other group in Fylingdales, indicating that this labouring population was the least stable group, with surnames turning over rapidly. While in the early decades their pattern echoes the farmers, they finally make a large move back to the centre of the plot.

Such movement towards the centre of the figure is a feature of three occupations: agricultural labourers, mariners and 'others'. The latter two converge gradually over the 50-year period, and such movement is to be interpreted as a breakdown of differentiation within the parish, as the differences between the surnames in the various occupational groups diminished. For example, the progress of the group 'other' suggests that men from seafaring and farming families were moving into the crafts and trades, and the final position of the agricultural labourers indicates a transfer of names from traditionally maritime families from the sea to the land.

There is much about this kind of analysis that is speculative, and corroboration of specific inferences can often be gained only from the detail of censuses, parish registers and other documentary sources. There is a value, however, in the broad summary view which the method affords, and in this example the surname distributions in nineteenth-century Fylingdales have indeed revealed a structuring of the population according to occupation, and have suggested the dynamics of change in that structure with changing economic circumstances.

The application of isonymy to historical questions

In this section I argue that the capacity of random isonymy methods such as Lasker's R_i to reveal and quantify relationships and subdivisions of population suggests that they could find wider application to help resolve questions about population structure that have been formulated within historical debate. Since the inference from surnames is essentially as a surrogate of kinship, it follows that temporal, spatial or sociocultural correlates of kinship can be readily addressed, as in the case studies above.

In so far as surname distributions subsume the history of migration and mate choice, isonymy also provides a measure of outcome of the demographic processes and the economic and sociocultural factors which influence them. Thus, more general questions as to the effects on population structure of systems of land tenure and inheritance, enclosure, industrial development, epidemic disease or natural disaster might all be addressed in terms of the trace they leave on the distribution of surnames. To cite a concrete example, the debate on open and close parishes in England (Banks 1988; Mills 1959; Short 1992; Spencer 2000) could be illuminated by prediction and observation of the effect on surname distributions consequent on Mills's formulation of the demographic and structural consequences of varying responses to the Poor Laws.

The case studies used in this chapter to illustrate the method have exploited the demonstration and inference from within-parish diversity, but perhaps the greater significance of methods based on isonymy is that they can broaden the scale of resolution that is tractable to analysis, so that detailed studies at a level much greater than the parish can be undertaken. Since much of the research into kinship in the mainstream of historical demography has been based on studies using the labour-intensive method of family reconstitution, research has usually been constrained to the study of single parishes (e.g. Levine and Wrightson 1991; Wrightson and Levine 1979). Where the results of many studies have been amalgamated, inference is still on the basis of a number of single-parish studies (Wrigley *et al.* 1997). There is thus a limitation on scale imposed by the method, and while family reconstitution has clearly demonstrated for the early-modern period that many families did not reside in the same parish throughout the family life cycle, there is reason to believe that when families moved out of a parish, they did not necessarily move very far away. Apparent mobility at a parish level may thus mask stability at a larger local scale.

Since surname distribution studies are not limited to a single parish, one kind of problem which might be particularly amenable to analysis by isonymy is the question of scale of interaction between individuals and among populations, and the delineation of regions and their boundaries. This issue is one of the key debates in the academic study of local and regional history, and is also informed by social anthropologists' analysis of community (Strathern 1981, 1982; Cohen 1982) and the sociologist Giddens's concept of *locale* (Giddens 1984). Contributors to this debate have urged the recognition of a wider community of interaction than the parish, and argued for the enduring stability and identity of such larger-scale local populations. Terms such as 'farming region' (Thirsk 1967), 'neighbour-hood' or 'social area' (Spufford 1974), '*pays*' (Phythian-Adams 1987) or 'country' (Hey 2000) have been employed to express the cultural, farming and/or landscape region which is seen to be the locus of such identity. Others have emphasised the notion of 'community', and the concept of 'community history' has in turn been thoroughly scrutinised by Marshall (1997), another champion of the recognition of the predictive value of regional identity.

Phythian-Adams has carried this debate forward with his argument for the recognition a general scheme of larger-scale interaction than that conventionally represented by the parish. In an earlier work (Phythian-Adams 1987) he advanced and offers a test of the claim that administrative counties might mark boundaries which constrain mobility. Subsequently (Phythian-Adams 1994) he has provided a valuable practical initiative by using landscape features such as coasts, watersheds and river drainage patterns to designate 'cultural provinces' throughout England. The boundaries between these are often coincident with some but not all of the old (pre-1974) administrative county boundaries in England.

My point here is not to advance or test the claims made for these drainage regions as actual correlates of cultural regions, but rather to suggest that this framework, and of course alternatives, might be explored in a rigorous hypothesis-testing way by analysis of surname distributions. If it were the case that the parish, say, was the principal arena for social interaction, including mate choice and mobility, then we should predict homogeneity within parishes and heterogeneity between them. If, however, landscape features as mentioned above delineate wider social boundaries, then it is at these features that we should expect greater discontinuities in surname distributions, with between-parish differences predicted to be of less importance. To give a specific example, Phythian-Adams's region 8 (Thames Estuary) adjoins region 6 ('French' Channel) (Phythian-Adams 1994). Their common boundary is the border of Kent with Sussex. Region 6 continues to the west including the counties of Sussex, Hampshire, Wiltshire and Dorset. We could compare surname distributions across all these county boundaries to see whether the one which separated two of the proposed cultural provinces marked a greater change in surname distributions than those county boundaries which are within the 'French Channel' region.

The kinds of questions suggested in this section – kinship, inheritance, enclosure, industrialisation and the delineation of boundaries between cultural regions – require the ability to analyse information at different levels of resolution. Surname distributions permit this, being relatively easy to collect and analyse, and through the method of isonymy offer a quantitative approach to the evaluation of hypotheses based on other premises and data.

I offer this as a case where a technique developed and routinely applied in one field may yet find a place as a tool for the solution of problems having their origin in a different area of discourse. Conversely, of course, the interpretation of patterns of genetic variation, which motivates the study of surname distributions in biological anthropology, can benefit greatly from a more informed appreciation of historical processes.

References

Banks, S. (1988) 'Nineteenth-century scandal or twentieth-century model? A new look at "open" and "close" parishes', *Economic History Review*, 61, 51–73.

Barrai, I., Scapoli, C., Beretta, M., Nesti, C., Mamolini, E. and Rodriguez-Larralde, A. (1996) 'Isonymy and the genetic structure of Switzerland. 1. The distributions of surnames', *Annals of Human Biology*, 23, 431–55.

Barrai, I., Rodriguez-Larralde, A., Mamolini, E. and Scapoli, C. (1999) 'Isonymy and isolation by distance in Italy', *Human Biology*, 71, 947–61.

Barrai, I., Rodriguez-Larralde, A., Mamolini, E., Manni, F. and Scapoli, G. (2001) 'Isonymy structure of USA population', *American Journal of Physical Anthropology*, 114, 109–23.

Bittles, A. H. (1994) 'The role and significance of consanguinity as a demographic variable', *Population and Development Review*, 20, 561–84.

Bittles, A. H. (1998) *Empirical Estimates of the Global Prevalence of Consanguineous Marriage in Contemporary Societies*, working paper no. 0074, Stanford, CA: Morrison Institute for Population and Resource Studies, Stanford University.

Bittles, A. H. and Neel, J. V. (1994) 'The costs of human inbreeding and their implications for variations at the DNA level', *Nature Genetics*, 8, 117–21.

Boldsen, J. L., Mascie-Taylor, C. G. N. and Lasker, G. W. (1986) 'An analysis of the geographical distribution of selected British surnames', *Human Biology*, 58, 85–96.

Bramwell, B. S. (1939) 'Frequency of cousin marriages', *Genealogists' Magazine*, 8, 305–16.

Buckatzsch, E. J. (1951) 'The constancy of local populations and migration before 1800', *Population Studies*, 5, 62–9.

Cavalli-Sforza, L. L. and Bodmer, W. F. (1971) *The Genetics of Human Populations*, San Francisco, CA: Freeman.

Chakraborty, R. (1985) 'A note on the calculation of random RP and its sampling variance', *Human Biology*, 57, 713–17. (Erratum, *Human Biology*, 58, 991.)

Christensen, A. F. (1999) 'Population relationships by isonymy in frontier Pennsylvania', *Human Biology*, 71, 859–73.

Cohen, A. P. (1982) 'Belonging: the experience of culture', in Cohen, A. P. (ed.) *Belonging*, Manchester: Manchester University Press, pp. 1–17.

Coleman, D. A. (1984) 'Marital choice and geographical mobility', in Boyce, A. J. (ed.) *Migration and Mobility*, London: Taylor & Francis, pp. 19–55.

Crow, J. F. (1980) 'The estimation of inbreeding from isonymy', *Human Biology*, 52, 1–14.

Crow, J. F. (1989) 'The estimation of inbreeding from isonymy – update', *Human Biology*, 61, 949–54.

Crow, J. F. and Mange, A. P. (1965) 'Measurement of inbreeding from the frequency of marriages of persons of the same surname', *Eugenics Quarterly*, 12, 199–203.

Darwin, G. H. (1875) 'Marriages between first cousins and their effects', *Journal of the Statistical Society*, 38, 153–84.

Giddens, A. (1984) *The Constitution of Society*, Cambridge: Cambridge University Press.

Giggs, J. A. (1994) 'Surname geography: a study of the Giggs family name 1450–1989', *East Midland Geographer*, 17, 58–67.

Gottlieb, K. (ed.) (1983) 'Surnames as markers of inbreeding and migration', *Human Biology*, 55, 209–408.

Guppy, H. B. (1890) *Homes of Family Names in Great Britain*, London: Harrison & Sons.

Halse, B. (2000) 'Population mobility in the village of Levisham, 1541–1900', *Local Population Studies*, 65, 58–63.

Hanks, P. (1992–3) 'The present-day distribution of surnames in the British Isles', *Nomina*, 16, 79–98.

Hanks, P. and Hodges, F. (1988) *A Dictionary of Surnames*, Oxford: Oxford University Press.

Heron-Allan, E. (1911) *Selsey Bill: historic and prehistoric*, London: Duckworth.

Hey, D. (1993) *The Oxford Guide to Family History*, Oxford: Oxford University Press.

Hey, D. (1997) 'The local history of family names', *Local Historian*, 27, 1–20.

Hey, D. (2000) *Family Names and Family History*, London: Hambledon & London.

Hey, D. (2001) 'Surnames, genes and genealogy: beyond the broadcast factsheet', http://www.bbc.co.uk/education/beyond/factsheets/surnames/surnames_home.shtml

Hill, E. W., Jobling, M. A. and Bradley, D. G. (2000) 'Y-chromosome variation and Irish origins', *Nature*, 404, 351–2.

Holloway, S. M. and Sofaer, J. A. (1989) 'Coefficients of relationship by isonymy within and between the regions of Scotland', *Human Biology*, 61, 87–97.

Hoskins, W. G. (1972) *Local History in England*, 2nd edn, London: Longman, Chapter 11, 'The homes of family names'.

Howell, C. (1983) *Land, Family and Inheritance in Transition: Kibworth Harcourt 1280–1700*, Cambridge: Cambridge University Press.

Huth, A. H. (1879) 'An index to books and papers on marriage between near kin', *Appendix to the Report of the Index Society*. London.

Jobling, M. A. (2001) 'In the name of the father: surnames and genetics', *Trends in Genetics*, 17, 353–7.

Koertvelyessy, T., Crawford, M. H., Pap, M. and Szilagyi, K. (1992) 'The influence of religious affiliation on surname repetition in marriages in Tiszaszalka', *Hungary Journal of Biosocial Science*, 24, 113–21.

Kosten, M. and Mitchell, R. J. (1990) 'Examining population-structure through the use of surname matrices: methodology for visualizing nonrandom mating', *Human Biology*, 62, 319–35.

Lasker, G. W. (1977) 'A coefficient of relationship by isonymy: a method for estimating the genetic relationship between populations', *Human Biology*, 49, 489–93.

Lasker, G. W. (1985) *Surnames and Genetic Structure*, Cambridge: Cambridge University Press.

Lasker, G. W. (1997) 'Census versus sample data in isonymy studies: relationships at short distances', *Human Biology*, 69, 733–8.

Lasker, G. W. (1999) 'The hierarchical genetic structure of an urban town, Kidlington, Oxfordshire, examined by the coefficient of relationship by isonymy', *Journal of Biosocial Science*, 31, 279–84.

Lasker, G. W. and Kaplan, B. A. (1985) 'Surnames and genetic structure: repetition of the same pairs of names of married couples, a measure of subdivision of the population', *Human Biology*, 57, 431–40.

Lasker, G. W. and Mascie-Taylor, G. C. N. (1985) 'The geographical-distribution of selected surnames in Britain: model gene-frequency clines', *Journal of Human Evolution*, 14, 385–92.

Lasker, G. W., Mascie-Taylor, C. G. N. and Coleman, D. A. (1986) 'Repeating pairs of surnames in marriages in Reading (England) and their significance for population structure', *Human Biology*, 58, 421–6.

Leland, J. (1964) *The Itinerary of John Leland in or about the Years 1535–1543*, edited by Lucy Toulmin Smith. London: Centaur Press.

Levine, D. and Wrightson, K. (1991) *The Making of an Industrial Society: Whickham 1560–1765*, Oxford: Clarendon.

McKinley, R. A. (1969) *Norfolk Surnames in the Sixteenth Century*, Leicester: Leicester University Press.

McKinley, R. A. (1975) *Norfolk and Suffolk Surnames in the Middle Ages*, London: Phillimore.

McKinley, R. A. (1977) *The Surnames of Oxfordshire*, London: Leopard's Head Press.

McKinley, R. A. (1981) *The Surnames of Lancashire*, London: Leopard's Head Press.

McKinley, R. A. (1988) *The Surnames of Sussex*, Oxford: Leopard's Head Press.

Marshall, J. (1997) *The Tyranny of the Discrete: a discussion of the problems of local history in England*, Aldershot: Scolar Press.

Mascie-Taylor, C. G. N., Lasker, G. W. and Boyce, A. J. (1987) 'Repetition of the same surnames in different marriages as an indication of the structure of the population of Sanday Island, Orkney Islands', *Human Biology*, 59, 97–102.

Mills, D. R. (1959) 'The Poor Laws and the distribution of population, c. 1600–1860, with special reference to Lincolnshire', *Transactions of the Institute of British Geographers*, 26, 185–95.

Morton, N. E. (1973) *The Genetic Structure of Populations*, Honolulu, HI: University of Hawai'i Press.

Morton, N. E. (1982) *Outline of Genetic Epidemiology*, Basel: Karger, Chapter 7, 'Population structure'.

Morton, N. E., Yee, S., Harris, D. E. and Lew, R. (1971) 'Bioassay of kinship', *Theoretical Population Biology*, 2, 507–24.

Phythian-Adams, C. (1987) *Rethinking English Local History*, Leicester: Leicester University Press.

Phythian-Adams, C. (1994) 'An agenda for English local history', in Phythian-Adams, C. (ed.) *Societies, Cultures and Kinship, 1580–1850*, London: Leicester University Press, pp. 1–23.

Pinto-Cisternas, J., Pineda, L. and Barrai, I. (1985) 'Estimation of inbreeding by isonymy in Iberoamerican populations: an extension of the method of Crow and Mange', *American Journal of Human Genetics*, 37, 373–85.

Pollitzer, W. S., Smith, M. T. and Williams, W. R. (1988) 'A study of isonymic relationships in Fylingdales parish from marriage records from 1654 through 1916', *Human Biology*, 60, 363–82.

Poole, O. (2001) 'Why the Dysons keep faith in their genes', *Sunday Telegraph*, 10 June. http://www.lineone.net/telegraph/2001/06/10/news/why_40.html

Porteous, J. D. (1982) 'Surname geography: a study of the Mell family name c. 1538–1980', *Transactions of the Institute of British Geographers*, ns 7, 395–418.

Postles, D. (1995) *The Surnames of Devon*, Oxford: Leopard's Head Press.

Postles, D. (1998) *The Surnames of Leicestershire and Rutland*, Oxford: Leopard's Head Press.

Reaney, P. H. (1967) *The Origin of English Surnames*, London, Routledge & Kegan Paul.

Redmonds, G. (1973) *Yorkshire, West Riding*. English surnames series, 1. Chichester: Phillimore.

Redmonds, G. (2001) *Surnames, Genes and Genealogy*. BBC Radio 4.

Relethford, J. H. (1988) 'Estimation of kinship and genetic distance from surnames', *Human Biology*, 60, 475–92.

Relethford, J. H. (1992) 'Analysis of marital structure in Massachusetts using repeating pairs of surnames', *Human Biology*, 64(1), 25–33.

Rodriguez-Larralde, A. and Barrai, I. (1997) 'Isonymy structure of Sucre and Tachira, two Venezuelan states', *Human Biology*, 69, 715–31.

Rodriguez-Larralde, A., Barrai, I. and Alfonzo, J. C. (1993) 'Isonymy structure of 4 Venezuelan states', *Annals of Human Biology*, 20, 131–45.

Rodriguez-Larralde, A., Barrai, I., Nesti, C., Mamolini, E. and Scapoli, C. (1998a) 'Isonymy and isolation by distance in Germany', *Human Biology*, 70, 1041–56.

Rodriguez-Larralde, A., Scapoli, C., Beretta, M., Nesti, C., Mamolini, E. and Barrai, I. (1998b) 'Isonymy and the genetic structure of Switzerland. II. Isolation by distance', *Annals of Human Biology*, 25, 533–40.

Rodriguez-Larralde, A., Morales, J. and Barrai, I. (2000) 'Surname frequency and the isonymy structure of Venezuela', *American Journal of Human Biology*, 12, 352–62.

Rogers, C. D. (1995) *The Surname Detective*, Manchester: Manchester University Press.

Rowlands, J. and Rowlands, S. (1996) *The Surnames of Wales for Family Historians and Others*, Baltimore, MD: Genealogical Publishing.

Short, B. M. (1992) 'The evolution of contrasting communities within rural England', in Short, B. M. (ed.) *The English Rural Community: image and analysis*, Cambridge: Cambridge University Press, pp. 19–43.

Skorecki, K., Sellg, S., Blazer, S., Bradman, R., Bradman, N., Waburton, P. J., Ismajlowicz, M. and Hammer, M. F. (1997) 'Y chromosomes of Jewish priests', *Nature*, 385, 32.

Smith, D. S. (1989) '"All in some degree related to each other": a demographic and comparative resolution of the anomaly of New England', *American Historical Review*, 94, 44–79.

Smith, M. T. and Hudson, B. L. (1984) 'Isonymic relationships in Fylingdales parish, North Yorkshire, in 1851', *Annals of Human Biology*, 11, 141–8.

Smith, M. T., Williams, W. R., McHugh, J. J. and Bittles, A. H. (1990) 'Isonymic analysis of post-Famine relationships in the Ards Peninsula, N. E. Ireland: genetic effects of geographical and politico-religious boundaries', *American Journal of Human Biology*, 2, 245–54.

Souden, D. and Lasker, G. W. (1978) 'Biological inter-relationships between parishes in east Kent: an analysis of Marriage Duty Act returns for 1705', *Local Population Studies*, 21, 30–9.

Spencer, D. (2000) 'Reformulating the "closed" parish thesis: associations, interests and interaction', *Journal of Historical Geography*, 26, 83–98.

Spufford, M. (1974) 'Contrasting communities: English villagers in the sixteenth and seventeenth centuries', Cambridge: Cambridge University Press.

Stapleton, B. (1999) 'Family strategies: patterns of inheritance in Odiham, Hampshire, 1525–1850', *Continuity and Change*, 14, 385–402.

Strathern, M. (1981) *Kinship at the Core: an anthropology of Elmdon, Essex*, Cambridge: Cambridge University Press.

Strathern, M. (1982) 'The place of kinship: kin, class and village status in Elmdon, Essex', in Cohen, A. P. (ed.) *Belonging*, Manchester: Manchester University Press, pp. 72–100.

Sykes, B. and Irven, C. (2000) 'Surnames and the Y chromosome', *American Journal of Human Genetics*, 66, 1417–19.

Thirsk, J. (1967) 'The farming regions of England', in Thirsk, J. (ed.) *The Agrarian History of England and Wales, Volume IV*, Cambridge: Cambridge University Press, pp. 1–112.

Thomas, M. G., Skorecki, K., Ben-Ami, H., Parfitt, T., Bradman, N. and Goldstein, D. B. (1998) 'Origins of Old Testament priests', *Nature*, 394, 138–40.

Walmsley, L. (1997) *The Sound of the Sea*, Otley: Smith Settle.

Watson, R. (1975) 'A study of surname distribution in a group of Cambridgeshire parishes 1538–1840', *Local Population Studies*, 15, 23–32.

Wolf, H. W. (1882) *Quaint Industries and Interesting Places in Sussex*, Lewes: Sussex Advertiser.

Wright, S. (1951) 'The genetical structure of populations', *Annals of Eugenics*, 15, 323–54.

Wrightson, K. and Levine, D. (1979) *Poverty and Piety in an English Village: Terling, 1525–1700*, New York: Academic Press.

Wrigley, E. A., Davies, R. S., Oeppen, J. E. and Schofield, R. S. (1997) *English Population History from Family Reconstitution, 1580–1837*, Cambridge: Cambridge University Press.

Wyatt, G. (1989) 'Population change and stability in a Cheshire parish during the eighteenth century', *Local Population Studies*, 43, 47–52.

Young, G. (1817) *A History of Whitby*, Whitby: Clark & Medd.

6 Calculating nutritional status in the past from historical sources

Derek J. Oddy

Introduction

During the 1950s, economic historians began to debate how the Industrial Revolution had affected the standard of living of the ordinary people in Britain.[1] In doing so, they necessarily had to consider the supply and prices of such mundane commodities as foodstuffs. Their work was concerned mainly with trends in the supply of food commodities derived from market data rather than food consumption in total. At the same time, economic historians were also becoming interested in demography and the study of population statistics – particularly deaths.[2] Examining the causes of deaths led to speculation about the presence of nutritional deficiency diseases in the population in the past. By these routes, food consumption came within the scope of historical analysis for the first time. Prior to this, the consumption of food had not been a subject for serious historical research but more a matter of antiquarian interest. The only sustained study of food consumption, *The Englishman's Food*, had not been written by historians but was a hobby interest of biochemists.[3] So, when Peter Laslett asked the question 'Did the peasants really starve?' in a BBC Third Programme broadcast in the early 1960s,[4] it signified not only the growing interest in food consumption among historians and social scientists but, in addition, an interest in the nutritional status that resulted from people's diets in the past. In 1966, John Burnett's book, *Plenty and Want*,[5] was the first volume to combine these questions within academic social history.[6] Traditional political historians continued to ignore food in their writings unless scurvy limited naval power or famine became so severe as to affect government policies. The lack of food, it seemed, particularly when it reached famine proportions, was of more interest to political historians than food itself. Most, however, saw these as subjects better left to specialists.[7]

The next step in the development of more analytical work, involving the nutritional analysis of food consumption data, depended upon quantification and the use of computers, although the available sources initially appeared unpromising. There were institutional records listing food supplied to schools, hospitals and workhouses, or ships' victualling accounts and army rations. By their nature, these were usually incomplete and unsuitable for the purposes of calculating energy and nutrient intakes. The occasional descriptions of specific events that survived, such as banquets or tithe dinners, seldom recorded either amounts of food or numbers of guests, so that quantification proved difficult or even meaningless.[8]

The value of family budgets as a source of information that might be used for nutritional analysis was recognized in the 1960s, when a Social Nutrition Research Unit was formed in the Department of Nutrition at Queen Elizabeth College, London.[9] Part of its programme was to examine dietary change during the period of industrialisation in Britain. Family budget data were central to this research project because family budgets reflected normal living conditions rather than prescribed diets found in institutional settings such as schools, hospitals, prisons or the armed services. Family budgets, however, do have their limitations as a form of evidence. They began to be collected only in the 1780s and the social surveys that provided them went out of fashion with the outbreak of the Second World War in 1939.[10] Moreover, the collectors of family budgets seldom intended them to provide dietary information. Usually they were concerned with general questions about living conditions; family budgets were frequently collected in connection with the poverty question, and so reflect for the most part life in the poorer parts of British society. Nevertheless, between the 1790s and 1940, family budgets are a major source of food consumption evidence for people living normal lives in domestic households and can be used to show trends in nutritional status.

Sources for calculating nutritional status

The family budgets contained within David Davies, *The Case of Labourers in Husbandry* (1795) and Sir Frederic Morton Eden, *The State of the Poor* (1797) are a case in point.[11] These books provide the largest source of data on food consumption by families exercising free choice in the early stages of industrialisation in Britain.[12] Both Davies and Eden indicate clearly the reason for their inquiries. Davies, dedicating his work to the Board of Agriculture, hoped he was 'drawing once more the attention of considerate persons to what appears to be a case of real, widespread, and increasing distress'. Similarly, Eden referred to his feelings of benevolence and personal curiosity about the condition of the labouring classes in view of the 'high price of grain and of provisions in general, as well as of cloathing and fuel, during the years 1794 and 1795'.[13] Their motives fitted contemporary opinion: Dorothy Marshall noted that the severity of attitude usual in the early eighteenth century, when it was felt that the poor should be confined to workhouses, had been modified in the later years of the century. 'Then, with the rise of prices after the '70's there came a gradual change of opinion as conditions began to throw some doubt on this hypothesis, and the general attitude became softened by a more sympathetic outlook.'[14] However, Dorothy Marshall felt that the typical contemporary explanations of poverty were still to be found either in the laziness or the luxury of the poor. Some criticism of the poor law regulations had already occurred and Davies asked that his inquiry should be taken into account in the framing of new regulations 'as these accounts form the groundwork of what I have to advance on behalf of the poor.'[15]

SIDLESHAM PARISH, SURRY

[COMMUNICATED BY JOHN FARHILL, ESQ; 1793]

	No. 1 6 Persons		
Expences per Week	£.	s.	d
Bread and Flour — — — — — — — — — —	0	4	0
Yeast and Salt — — — — — — — — — —	0	0	0½
Bacon and other Meat — — — — — — — —	0	2	0
Tea and Sugar — — — — — — — — — —	0	0	7
Butter — — — — — — — — — — — —	0	0	9
Cheese — — — — — — — — — — —	0	1	0
Soap, Starch, and Blue — — — — — — — —	0	0	2
Candles — — — — — — — — — — —	0	0	2½
Thread, Worsted, &c. — — — — — — — —	0	0	1
Total	0	8	10
Amount *per annum*	22	19	4
Earnings per Week	£.	s.	d
Total	0	10	0
Amount *per annum*	26	0	0
	£.	s.	d
Expences *per annum*	22	19	4
Rent and Fuel	3	13	6
Total Expences *per annum*	26	12	10
Total Earnings *per annum*	26	0	0
	0	12	10
			Deficient

ACCOUNT OF THE FAMILIES

No. 1. John Hart, his wife, and four children, the eldest a girl ten years old, another four, a boy two, and the youngest an infant. Earn about 10s. per week

ANNUAL EXPENCES

No. 1 Rent   1   2   6
Coals 1l. 10s.
Wood 1l. 1s.
                3   13   6
Brews 6 bushels of malt,
4lbs. Hops.

Table 6.1 A family budget collected by David Davies

Even the limited information available raises the question of how Davies and Eden set about their inquiries without a major disruption of their other activities. Fortunately, both are explicit on this point. Davies informed his readers that his inquiries originated in his parish visiting work where he found distress that could not be imputed to sloth or wastefulness:

'Every thing (said they) is so dear, that we can hardly live'. In order to assure myself, whether this was really the case, I enquired into the particulars of their earnings and expences; and wrote the same down at the time, just as I received them from each family respectively, guarding as well as I could against error and deception.[16]

Davies's initial inquiry was at Easter 1787, 'when affairs relating to the poor were under consideration of the Parliament and the public'. After some 'discussion', he apparently made an 'abstract' which he then circulated, though there is no further explanation of the process. His expectations were that he should receive two or three papers from every county and he records his regret that the greater number was not returned. Among the details of family budgets that Davies gave in his Appendix, he noted: 'The first paper in this collection is that which was circulated for the purpose of obtaining information.' This was his six accounts of Barkham families. It may be assumed that he distributed this 'abstract' in completed form since other returned papers occasionally compared the similarity of their local prices with those in Barkham. One return, from Sidlesham in Surrey, is set out in Table 6.1 with its original spelling.[17]

Eden claimed to have designed a questionnaire that followed 'the not very dissimilar Queries' used by Sir John Sinclair and another inquiry by Sir John Cullum.[18] Eden included questions on population, housing circumstances, occupations, prices of provisions and labour, religious groups, tithes, ale-houses, size of farms, types of tenure, rentals and tax levels, the extent of enclosure, the existence of friendly societies, and the maintenance of the poor. Eden also claimed to have visited several parishes himself, selecting 'persons the most likely to supply useful information, and the least likely to be misinformed, or to mislead', and to have obtained information from 'a few respectable clergymen and others'.[19] An example of his summary budgets is given in Table 6.2. Additionally, he seems to

DURHAM – St. NICHOLAS

   The following are the earnings and expences of a man who is an hostler at one of
   the inns in this city. He is 45 years of age; has 6 children, all boys; the eldest is 10
   years, and the youngest 9 months old.

EARNINGS

	£.	s.	d.
The man earns 9s. a week, (besides being allowed his diet;) yearly	23	8	0
His wife earns 2d. a week by spinning, yearly	0	8	8
Total earnings –	£ 23	16	8

EXPENCES

	£.	s.	d.
Barley meal, 3s. 4d. a week, yearly	8	13	4
Milk, 1s. 2d. a week, yearly	3	0	8
Potatoes, 8d. a week, yearly	1	14	8
Oatmeal, 10d. a week, yearly	2	3	4
Tea and sugar, 1s. a week, yearly	2	12	0
Soap, blue, &c. 3d. a week, yearly	0	13	0
Butcher's meat, 10d. a week, yearly	2	3	4
Salt, 1d. a week, yearly	0	4	4
House rent, yearly	1	0	0
Fuel, yearly	1	6	0
Lying-in costs annually, about	0	8	0
Cloaths, and other expences, yearly about	2	10	0
Total expences	£ 26	8	8

No butter or beer is used by this family: they occasionally receive
a few old cloaths from their neighbours; but do not ask relief of
the parish

March 1796

Table 6.2 A family budget collected by Sir Frederic Morton Eden

have used an investigator: 'To other parishes and districts, not thus accessible to me, I sent a remarkably faithful and intelligent person; who has spent more than a year in travelling from place to place, for the express purpose of obtaining exact information'.[20] Somewhat surprisingly, Eden made no reference to Davies's earlier inquiries nor his considerable debt to Davies for both inspiration and methodology. However, it can be established without question that Eden had seen Davies's Abstract and had used it to obtain six budgets from Glynd, Sussex, in 1793.[21]

The result of these investigations is that Davies's Appendix included 123 family budgets from England, 5 from Wales and 9 from Scotland, totalling 137 in all. They were collected over a period of several years between 1787 and 1793. Eden recorded 43 family budgets collected in England and 3 from Wales, totalling 46, which he obtained between 1793 and 1796.[22] The dietary analysis that follows comprises 119 of Davies's families and 32 of Eden's.[23]

Dietary analysis

The methodology employed in the process of dietary analysis attempts to quantify food consumption in terms of a number of standard food categories, such as bread, meat, fats and so on, whereby the terms used cover a variety of foods within that category. The nutrient content of the diet has been analysed using standard food tables.[24] Certain assumptions were required in the process. The information provided by Davies and Eden is usually given in terms of expenditure on certain foods either by week or month or year. Some quantities are given but mostly quantities have been obtained by using prices as divisors. If these were not available for any specific food, a standard price was employed which reflected the average of all prices given for that commodity. Table 6.3 gives examples.

The nutrient analysis depended on the assumption that all meals were eaten at home or were provided out of the stock of food materials mentioned in the budget. Agricultural labourers commonly took food with them for a midday meal. Therefore any budget specifically indicating that a man usually ate at work in the farmhouse (which might be expressed in the terms that he was paid so-much-a-day and diet) was discarded – as Table 6.2 was – unless it appeared that the practice was limited to a period such as harvest, and that the budget did not refer to such time. The total quantity of food was divided by the total number of persons in the household, even when this included small children or infants so that food consumption and nutrient values of the diet are always expressed in amounts per head. The alternative method of constructing consumption allowances for men, women and children of different ages has not been used.[25]

As standard food tables are now available in electronic book form,[26] a number of dietary analysis programmes enable this work to be done on desktop computers. Using modern food tables leads to two problems for the historian; it is essential first, to avoid fortified foods and second, to assess what qualitative changes in food composition have occurred in the period since the historical evidence was collected. In using historical material for a purpose for which it was not originally intended,

Table 6.3 Examples of standard prices adopted

Food	Number of prices	Average price	Standardized price
Beef	15	4.7*d*./lb	4.5*d*./lb
Mutton	15	4.9*d*./lb	5.0*d*./lb
Veal	7	5.6*d*./lb	5.5*d*./lb
Bacon	11	8.6*d*./lb	8.5*d*./lb
Butter	17	10.1*d*./lb	10.0*d*./lb
Milk	13	0.72*d*./pt	0.75*d*./pt
Cheese	4	6.25*d*./lb	6.0*d*./lb
Eggs	1	0.5*d*. each	0.5*d*. each
Sugar	2	9.0*d*./lb	9.0*d*./lb
Treacle	1	4.0*d*./lb	4.0*d*./lb
Bread	5	2.76*d*./lb	2.75*d*./lb[a]
Flour	3	2.29*d*./lb	2.25*d*./lb

Note: If bread prices were given instead of flour, it was necessary to covert bread to flour for analysis. Calculations based on contemporary documents suggest 1.39 is a suitable conversion factor to give the ratio of bread to flour. For confirmation, see Dr Irving's recipe for bread given to a Committee of the House of Commons, quoted in Eden, vol. 1, p. 533. It gave a conversion factor of 1.39. Dr Edward Smith's report on the Lancashire Cotton Famine gave a conversion factor of 1.4. See *Fifth Report of the Medical Officer of the Privy Council* (1863: 322).

it is possible that some errors will occur in the process of dietary analysis. In fact, the scale of error is likely to be small, even very small, when limited amounts of food have to be apportioned between several members of a family and, as is often the case, spread over a seven-day period.

Results

The analyses of Davies's and Eden's surveys presented in the following tables show food consumption in weekly amounts per head and the nutrient intake per head per day. Only the main food categories are shown; minor food items consumed such as fish, eggs and vegetables other than potatoes were included in the nutrient analysis but have been left out of the Weekly Food Consumption table. 'Sugar' is the total weight of sugar together with treacle or molasses and 'fats' is generally butter but includes dripping. 'Beer' refers to that amount which was consumed at home. Although the numbers in Tables 6.4, 6.5, 6.6 and 6.7 are small, an attempt has been made to show the geographical dispersion of the families surveyed and, by use of means for northern and southern areas of England and Wales, to show the differences otherwise concealed by overall averages. The division into 'north' and 'south' is based on the high-wage and low-wage areas shown in James Caird's 1850–1 study, even though this is not directly comparable with conditions at the end of the eighteenth century.[27]

The budgets in the Davies survey in Table 6.4 suggest that a marked regional variation in food consumption existed in the late eighteenth century. Families in the north and west ate a diet of around 7–8 lbs of bread per head per week. Oatmeal, potatoes and milk complemented the high bread consumption so that

Table 6.4 Dietary evidence analysed from family budgets: weekly food consumption per head showing regional variations 1787–93

Area	Number	Meat (lb)	Cheese (oz)	Milk (pt)	Fats (oz)	Potatoes (lb)	Bread (lb)	Oatmeal (oz)	Sugar (oz)	Beer (pt)
A	14	0.27	4.4	2.05	1.2	2.47	9.02	3.6	1.8	0
B	10	0.26	0.4	1.16	0.8	1.02	5.2	38.4	2.4	0.6
C	1	0.45	6.7	2.33	2.2	0	7.71	0	0	0
D	6	0.07	0.9	0.78	1.1	0	8.12	0	1.1	0
E	6	0.59	1	0	1.9	0	7.53	0	2	0
F	41	0.42	3.2	0	1.9	0	9.32	0	1.8	3.8
G	36	0.36	3.9	0.2	0.9	0.08	10.47	0	1.7	0.1
H	5	0	0	0.87	6	1.63	7.2	49.2	0	0
Mean	119	0.34	3.0	0.49	1.6	0.47	9.03	5.7	1.7	1.4
Mean A–C	25	0.27	2.9	1.71	1.1	1.79	7.44	17.4	2.0	0.2
Mean D–H	94	0.36	3.0	0.17	1.7	0.12	9.46	2.6	1.6	1.7

Key to areas: A Northern Counties; B Lancs/Yorks; C North Midlands; D West Midlands; E East Midlands; F South-East; G South-West; H Wales
Counties were allocated as follows: A = Cumberland, Durham, Northumberland, Westmorland; B = Cheshire, Lancashire, Yorkshire; C = Derby, Leicester (northern half), Lincoln, Nottingham, Shropshire (north-east), Staffordshire; D = Gloucester, Hereford, Oxford, Shropshire (south-west and south), Warwick, Worcester; E = Bedford, Cambridge, Huntingdon, Leicester (south), Northampton, Rutland; G = Berkshire, Buckingham, Hampshire, Hertford, Kent, Middlesex, Surrey, Sussex; G = Cornwall, Devon, Dorset, Somerset, Wiltshire; H = Denbigh, Merioneth. There were no usable budgets for East Anglia.

Source: David Davies (1795) *The Case of Labourers in Husbandry*

porridge and milk formed an important part of the diet. In the Midlands and southern districts of England bread consumption was over 9 lbs per head per week. There, it was the principal component of the diet, since no oatmeal was used and few potatoes eaten. More meat was eaten, so that the diets fell within the bread, bacon and beer pattern William Cobbett applauded in Sussex in 1830.[28]

The domestic economy of the agricultural family in southern England, depended, in Davies's view, upon having sufficient funds to obtain a pig from a farmer below market price, obtaining credit to buy flour by the sack, and having 'sufficient garden ground for planting a good patch of potatoes'.[29] Domestic brewing required families to use at least a bushel of malt every month to obtain 28 gallons of small beer, and the cost of materials and fuel (plus the cost of the equipment for young men setting up their own families) meant that the practice was declining.

For the poor in southern England, it was generally accepted that 'bread makes the principal part of the food of all poor families and almost the whole of the food of . . . large families'.[30] Many meals were made up of bread and tea alone, while the general demand for white, wheaten bread meant that barley and oatmeal, for use as porridge or 'hasty pudding', were in decline in the late

Table 6.5 Dietary evidence analysed from family budgets: daily nutrient intake per head showing regional variations, 1787–93

Area	Number	Energy (kcal)	Value (kj)	Protein (g)	Fat (g)	Carbohydrate (g)	Iron (mg)	Calcium (g)
A	14	2,029	8,489	58	29	384	12.9	0.47
B	10	1,866	7,807	49	33	344	11.2	0.28
C	1	1,803	7,544	50	45	300	6.5	0.40
D	6	1,543	6,456	38	15	314	6.4	0.19
E	6	1,534	6,418	43	49	292	7.5	0.11
F	41	1,994	8,343	46	33	379	7.6	0.22
G	36	2,103	8,799	51	32	401	7.0	0.21
H	5	2,459	10,288	60	46	452	11.4	0.24
Mean	119	1,992	8,336	49	33	378	8.4	0.25
Mean A–C	25	1,955	8,178	54	31	365	12.0	0.39
Mean D–H	94	2,002	8,378	48	33	382	7.5	0.21

Key to areas: A Northern Counties; B Lancs/Yorks; C North Midlands; D West Midlands; E East Midlands; F South-East; G South-West; H Wales

Source: David Davies (1795) *The Case of Labourers in Husbandry*

eighteenth century. Moreover, the price of fuel in southern England was already forcing poor people to buy bread from the baker rather than bake it themselves. The use of the potato in the south was limited until the nineteenth century because, as Davies explained:

> Wheaten bread maybe eaten alone with pleasure; but potatoes require either meat or milk to make them go down: you cannot make many hearty meals of them with salt and water only. Poor people indeed give them to their children in the greasy water in which they have boiled their greens and their morsel of bacon.[31]

Davies was not against potatoes, as he realised that they would help poor people to feed a pig, but felt rather that such dietary change was unlikely to be accepted in the south. The diets he records were not nutritionally adequate as the analysis in Table 6.5 shows. Daily energy values around 2,000 kcals (8,370 kj) were low for men engaged in manual work and also for women, especially when pregnant or lactating. Despite the differing amounts of food consumed, the 'northern' and 'southern' patterns of eating varied little in terms of energy values, fat and carbohydrate (CHO) intakes. Eating patterns in northern and western districts gave higher protein and iron intakes, but neither 'northern' nor 'southern' diets produced anything like adequate calcium intakes compared with modern recommended amounts.[32] The diets of the families surveyed by Davies were notable for their high carbohydrate content. Roughly three-quarters of the energy value came from carbohydrate sources. As the ratio of energy obtained from proteins varies little, it follows that these diets were extremely low in fat content, and low fat diets lack palatability.

Table 6.6 Dietary evidence analysed from family budgets: weekly food consumption per head showing regional variations, 1793–96

Area	Number	Meat (lb)	Cheese (oz)	Milk (pt)	Fats (oz)	Potatoes (lb)	Bread (lb)	Oatmeal (oz)	Sugar (oz)	Beer (pt)
A	16	0.55	0.5	4.77	3.5	6.90	3.92	41.7	2.5	0.8
B	2	1.60	8.6	2.66	11.1	9.00	3.81	34.3	14.0	0.8
C	1	1.80	5.4	1.33	5.4	4.19	2.23	0	2.7	2.8
D	3	0.16	0.7	0.58	1.6	2.40	8.14	0	1.4	0.3
E	3	1.04	5.0	1.70	3.0	3.14	6.42	0	3.2	2.2
F	3	0.70	3.3	0.11	2.6	0	7.44	0.4	4.7	0.3
G	1	0	2.7	0	4.2	4.05	14.03	0	0	0.0
H	3	0.17	0.3	0.37	2.0	2.93	8.34	0.8	0.1	0.0
Mean	32	0.63	1.9	2.85	3.6	5.06	5.55	23.1	3.1	0.8
Mean A–C	19	0.73	1.6	4.37	4.4	6.98	3.82	38.7	3.7	0.9
Mean D–H	13	0.48	2.4	0.64	2.4	2.27	8.08	0.3	2.2	0.7

Key to areas: A Northern Counties; B Lancs/Yorks; C North Midlands; D West Midlands; E East Midlands; F South-East; G South-West; H Wales

Source: Sir Frederic M. Eden (1797) *The State of the Poor*

Table 6.6 shows that food consumption by the families in Eden's survey was generally similar to those in the Davies survey. High bread consumption, at around 7 lbs per head per week, was noticeable in the south.[33] Potatoes were more commonly eaten in the north and west of Britain, as well as large amounts of oatmeal and milk. Sir Frederic Eden contrasted the southern diet of tea twice-a-day, best wheaten bread, and a little bacon with that of the north: 'In point of expense, their general diet as much exceeds, as, in point of nutrition, it falls short of, the north country fare of milk, potatoes, barley bread and hasty-pudding.'[34] Until the end of the eighteenth century, such a diet extended south into the Midlands and into Wales and the West Country, though there were already signs of change. At Burton-on Trent, Eden observed that 'oatmeal forms a great part of the food of the labouring classes: it is boiled with milk into a sort of hasty-pudding', though wheaten bread 'is now beginning to be introduced on particular occasions, by those who can afford it'.[35] The practice of farm servants living in was still widespread in the north of England until the nineteenth century, which meant that many ate at the farmer's table, though 'boarding out' became more common at times when food was dear.

Table 6.7, for Eden's families, shows somewhat higher energy values than those for Davies's families in Table 6.5. Eden's survey was undoubtedly weighted towards the northern counties of England and both Eden's and Davies's surveys agree that diets in the north yielded more energy and more of most nutrients than those in the south. The limited numbers make any more detailed comparison impossible. The fact that the surveys produce marked differences for families in Lancashire and Yorkshire, Wales and the South-West suggests that there can be

Table 6.7 Dietary evidence analysed from family budgets: daily nutrient intake per head showing regional variations, 1793–96

Area	Number	Energy (kcal)	Value (kj)	Protein (g)	Fat (g)	Carbohydrate (g)	Iron (mg)	Calcium (g)
A	16	2,352	9,841	70	53	397	14.4	0.65
B	2	2,986	12,493	92	94	442	18.0	0.69
C	1	1,636	6,845	59	58	217	9.9	0.37
D	3	1,729	7,234	43	17	351	7.9	0.18
E	3	1,889	7,904	58	33	341	9.7	0.41
F	3	1,689	7,067	46	33	300	7.5	0.18
G	1	2,919	12,213	64	28	603	6.4	0.16
H	3	1,798	7,523	42	17	370	5.2	0.11
Mean	32	2,171	9,084	62	44	379	11.7	0.47
Mean A–C	19	2,381	9,962	72	58	392	14.5	0.64
Mean D–H	13	1,864	7,800	49	25	361	7.5	0.22

Key to areas: A Northern Counties; B Lancs/Yorks; C North Midlands; D West Midlands; E East Midlands; F South-East; G South-West; H Wales

Source: Sir Frederic M. Eden (1797) *The State of the Poor*

different eating patterns in any region. Oatmeal appears to be important and its presence or lack of it in the diet explains the marked variation in energy levels in Lancashire and Yorkshire, and Wales. On the other hand, in the South-West the variation can be explained by the extraordinary amounts of bread consumed in both surveys. The diets in Tables 6.6 and 6.7 have similar characteristics to those in Tables 6.4 and 6.5, with energy values derived from low fat and high carbohydrate sources. In the midland and southern districts both surveys have very similar characteristics but Eden's northern families were beginning to obtain more fats in their diet and, in consequence, the proportion of energy derived from carbohydrates had fallen.

Long-term trends

Even at the time of Davies's and Eden's surveys, the domestic economy of the countryside was in transition. From the late eighteenth century brewing, as well as baking, began to decline in many rural areas. 'How wasteful, and indeed how shameful for a labourer's wife to go to the baker's shop' thundered William Cobbett,[36] but the decline in traditional self-sufficiency continued and from the mid-nineteenth century only pockets of resistance to shop products remained. Cooking and meal preparation took on a more limited form in rural districts in the nineteenth century and was mirrored in the newly urbanised areas by the disappearance of homemade soups, broths and pottages where these had formerly been traditional. Once this pattern of change is understood, it explains Maud Davies's comments in her 1906 survey of Corsley that wives cooked only once or twice a week during the winter, but perhaps more often when potatoes were plentiful, though 'it was exceptional to find a woman who . . . cooks every day.'[37] Indeed, the fact that people were satisfied to eat bread with cold meat or cheese and beer as regular everyday fare in much of England may well have limited the growth of culinary arts in Britain as much as any other factor.

One shortcoming of all historical evidence, whether from family budgets or supply estimates, is that the distribution of food within the family is seldom revealed. Yet circumstantial evidence points to a marked imbalance in food consumption within the family which must be understood for Tables 6.4 and 6.6 to be interpreted correctly. For example, in his report on the survey he made in 1863, Dr Edward Smith wrote: 'The remark was constantly made to me, "that the husband wins the bread and must have the best food".'[38] Because contemporaries accepted priorities of this kind as part of the normal patterns of behaviour in families where manual labour formed the basis of employment, such variation in food consumption was never measured and seldom commented on, though Smith concluded:

> The important practical fact is however well established, that the labourer eats meat and bacon almost daily, whilst his wife and children may eat it but once a week, and that both himself and his household believe that course to be necessary, to enable him to perform his labour.

Dr Thomas Oliver's study of food consumption in the 1890s is therefore interesting,[39] not only because it surveyed men like shepherds and navvies who frequently took their meals away from their families, but also because it included women who were the principal breadwinner in their families. Unfortunately, his sample was very small but, as shown in Table 6.8, there was a marked difference in food consumption by sex. The men ate far more meat, milk and potatoes than women, who in turn ate more sugar and fats than men. In effect, the men were eating cooked meals of meat and potatoes (and also porridge, which is why their milk consumption was so high) that provided adequately for their needs. On the other hand, the women, in spite of working in lead factories and cotton mills, largely confined themselves to eating bread with jam or treacle and drinking tea. This self-limitation of women's diet was apparent at times of economic crisis, as during the Lancashire Cotton Famine of the early 1860s,[40] but was almost certainly a permanent feature of life in lower-income families. In the early 1900s, Mrs Pember Reeves noted it in south London, as did Maud Davies in her study of Corsley in Wiltshire. In York, Rowntree's summarised it as:

> We see that many a labourer, who has a wife and three or four children, is healthy and a good worker, although he earns only a pound a week. What we do not see is that in order to give him enough food, mother and children habitually go short, for the mother knows that all depends upon the wages of her husband.[41]

Oliver offered a more sophisticated evaluation of the adequacy of the diets he had collected. The physiologists of the 1890s emphasised the importance of

Table 6.8 Weekly food consumption per head differentiated by sex, *c.*1894

	Number	Bread (lb)	Potatoes (lb)	Sugar (oz)	Fats (oz)	Meat (lb)	Milk (pt)
Men	21	3.4	4.6	15	5	4.1	2.0
Women	6	3.3	1.4	21	6	1.0	0

protein in the diet. They thought a man weighing 70 kilograms (11 stone) required 140 g protein a day on the basis that 2 g protein were required for each kilogram of body weight. Oliver himself thought this excessive but recommended that a workman's diet should include 20 g nitrogen per day, which was equivalent to 125 g protein.[42] In consequence, Oliver concluded with regard to the diets he published, 'with few exceptions, the individuals are not taking in their food that excess of proteid or animal food necessary to maintain their nitrogenous equilibrium'.[43] The analysis in Table 6.9 indicates that the average nitrogen content of Oliver's diets was 14.9 g (equivalent to 93 g of protein). Taking the sexes separately, men still fell short of his target of 120 g protein, while women achieved only 40 per cent of that figure. In view of the fact that his assessment was based on individual diets, it is not surprising that per capita intake of protein among families in Davies's and Eden's surveys was so far below Oliver's.

Table 6.9 Daily nutrient intake per head differentiated by sex, *c.*1894

	Energy (kcal)	Value (kJ)	Protein (g)	Fat (g)	Carbohydrate (g)	Iron (mg)	Calcium (g)
Men	3,321	13,895	114	146	387	19.0	0.69
Women	1,870	7,824	48	51	310	7.9	0.18

Given the emphasis placed on protein by physiologists in the 1890s, it was hardly surprising that the first attempts to justify a low-protein diet occurred at the end of the nineteenth century, even though there were signs that workloads were rising.[44] Moreover, as threats of war arose at the turn of the century, concern about the diet of the working classes became a strategic consideration.[45] Sivén, a Swedish physiologist, demonstrated in 1901 the possibility of living on a diet containing only 25 to 30 g protein per day, at least for a brief period.[46] In 1904, R. H. Chittenden of Yale spent a period of nine months on a diet providing on average 36 g protein per day.[47] As a result of further experiments on university teachers, soldiers and university students, he recommended that 60 g protein were sufficient for a 70 kg (11 stone) man.[48] Although Chittenden saw difficulties in obtaining support for his ideas in view of the prevailing opinions that 'by hearty eating lies the road to health and strength', nevertheless he envisaged greater efficiency of workers as a major and beneficial result.[49] These experiments were widely noted and led to some debate in the medical press in Britain. Reviewing

Chittenden's *Physiological Economy of Nutrition*, the *British Medical Journal* took the opportunity, with obvious relief, to belittle Rowntree's conclusions:

> Further, these results throw a flood of light upon what seemed the paradoxical statements of Mr. Seebohm Rowntree, whose philanthropic investigation into the lives of the poorer classes was published in his well-known book on *Poverty*. His statistics have been quoted by both sides on political platforms to show the enormous proportion of the working classes in this country who are underfed, but no one who knew these classes as the medical profession does could help doubting whether the figures quoted did not conceal some fallacy. The truth is that Mr. Rowntree accepted the current estimate of 125 grams of proteid as the amount necessary for a labouring man; and when he found, as he did in many cases, that the diet contained 90 grams or less, he naturally concluded that the people were underfed; but in the light of Professor Chittenden's results, the amount of proteid in the diets of Mr. Rowntree's working people was generally more than enough, being usually considerably above, and in only a few cases below, the 50 grams which have been proved to be sufficient for the subject of these experiments.[50]

However, a more recent analysis of Chittenden's low-protein diet of 36 to 40 g per day has pointed to its low energy value and deficiency of nutrients which should prevent it being considered adequate as a yardstick for normal life.[51] Thus the difficulties of using contemporary assessments of dietary standards are complicated by the conflicting viewpoints held by nutritionists and physiologists of the day. But, even with the recognition of these pitfalls, the problems of assessment are complicated further by the question of how to evaluate non-quantifiable factors affecting the diet. The essential difficulty is that diets expressed in aggregate terms cannot be related to the individual needs. No family budget survey gave detailed information about the individual members of the families examined.[52] The age of each person is given in most investigations, but no evidence whatsoever is available about individual body size and weight for a whole family, nor of special requirements by individuals due to the level of physical activity, to childhood, or to the later stages of pregnancy and lactation. Above all, it is a well-established fact that people show a remarkable degree of individual variation in their needs, regardless of their size, sex and level of physical activity.[53] A further problem that remains is the difficulty of defining an optimum state of health either for individuals or communities.[54] The historian who seeks to use dietary evidence in a discussion of the standard of living is therefore unlikely to obtain any firm assessment of the nutritional status of working-class families as a result of family budget surveys, unless there is corroboration from accompanying clinical studies.[55]

Nevertheless, family budget surveys continued to be collected at various intervals from the 1790s until the outbreak of the Second World War, and a long-term comparison of them provides the only way of showing how the structure of the diet in Britain changed. Figure 6.1 presents the energy composition of diets from the 1790s to 1937, a period covering the Industrial Revolution during

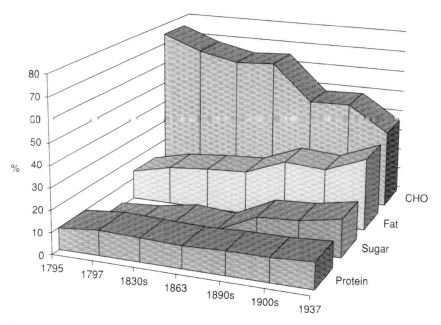

Figure 6.1 Energy sources of the British diet, 1790–1937

the nineteenth century and subsequent development of Britain as an urban-industrial society. Figure 6.1 breaks down the mean energy value of the diets into its component sources of proteins, fats and carbohydrates. To emphasise change, sugars are shown separately from the total of all carbohydrates (which includes sugars). The trend for proteins revealed by diets in family budget surveys shows little variation. Proteins provided around 10–11 per cent of the energy value of the diets, which differs little from the present day. Carbohydrates in total provided most of the energy throughout the period but, while this amounted to 65–75 per cent during the Industrial Revolution, the fall that set in from the 1860s onwards meant that by the 1930s only 36 per cent of energy came from carbohydrates. Conversely, sugars and fats grew in importance as sources of energy. Separating sugars from other carbohydrates reveals that sugars provided only 1.5–2.5 per cent of energy in the 1790s and no more than 4–6 per cent during the first half of the nineteenth century. From the 1860s, as free trade developed, sugars provided 13–14 per cent of energy at the end of the nineteenth and beginning of the twentieth centuries, and almost 18 per cent of energy by the late 1930s. Similarly, fats, which provided 18–20 per cent of energy during the Industrial Revolution period of the first half of the nineteenth century, also increased as free trade brought in more foodstuffs from North America and the southern hemisphere. By the late nineteenth and early twentieth centuries, some 27–8 per cent of energy in the diet came from fats and by the late 1930s this had reached 34 per cent. The rationing scheme of the 1940s delayed further changes in the composition of the diet for a decade. Rationing led to state monitoring of food consumption and with the growth of government social surveys that accompanied the Second World War came the end of the era of collecting family budgets.

Notes

1 The main debate occurred in the pages of the *Economic History Review* between 1957 and 1963.
2 A chronology of the growing interest in demography can be found in Drake (1969: 196–200).
3 Drummond and Wilbraham (1957).
4 Laslett (1971: Chapter 5; 1965: p. vii, Introduction). A similar question was posed by Williams (1976).
5 Burnett (1966).
6 George Fussell's pioneering attempt in the 1920s to evaluate changes in agricultural labourers' diet has largely been forgotten (see Fussell 1927).
7 This is not the place for a bibliography of work on famines, but famine has proved a subject of particular interest to historians with a nationalist agenda; for scurvy, see Watt *et al.* (1981).
8 A good example of a tithe dinner can be found in Parson Woodforde's diary. See James Woodforde (1978: 513–14) for a tithe dinner held on 1 December 1795.
9 Now reabsorbed into King's College London.
10 The government's War-Time Food Survey commenced regular nutritional inquiries in 1940, which were continued by the National Food Survey in the postwar period.
11 Neither the Reverend David Davies nor Sir Frederic Eden is particularly well documented. Davies does not appear to have written any other pamphlets on the subject of poverty and the paucity of information in his entry in the *Dictionary of National Biography* is of little help. He was at Jesus College, Oxford before being appointed Rector of Barkham in Berkshire, a post which he apparently held until his death about 1819. There is only a little more about Eden: he was born in 1766, possibly at Durham, the eldest son of Sir Robert Eden. He was at Christ Church, Oxford, but was not a contemporary of Davies, who took his first degree nine years earlier. Eden spent much of his life in commerce in London, being a founder and later chairman of the Globe Insurance Company, at whose office he died suddenly in 1809. Eden's London address was Lincoln's Inn Fields but he is buried at Ealing.
12 A comparison might be made with Fussell (1927).
13 Davies (1795: 3); Eden, (1797: vol. 1, Preface, p. i).
14 Marshall (1926: 14, 21–7).
15 See Marshall (1926) for a discussion of contemporary opinion about the poor, especially pp. 52–3; Davies (1795: 4).
16 Davies (1795: 6).
17 See Davies (1795: 180–1).
18 Eden (1797: Preface, pp. iv–v).
19 This raises the question of whether those he approached were capable of averaging their expenditure. In the 1860s, Dr Edward Smith noted that not everyone he surveyed could do so, particularly if food purchases were infrequent or irregular. See *Sixth Report of the Medical Officer of the Privy Council* (1864).
20 Eden (1797: Preface, p. ii).
21 Eden (1797: vol. 1, p. 734).
22 Eden also described the diet in 87 workhouses, all but 3 of which were in England. The descriptions vary considerably in detail but the dietary information from 23 or 24 of them might be analysed.

23 The analysis of food consumption based on the Davies and Eden material published by Shammas (1990: 134) used only 22 budgets.
24 McCance and Widdowson (1960) provided the database for the analysis of these budgets, which was carried out on the University of London mainframe computer.
25 This was commonly done in the interwar work of the Medical Research Council but from the Second World War became superseded by per capita allowances. See British Medical Association Committee on Nutrition (1950: 19–20, 29 *et seq.*). Consumption allowances have become popular again in recent years among anthropometric historians seeking spurious accuracy from inadequate data. See Shammas (1990: 135 *et seq.*) for calculations of energy values from Davies's and Eden's material using consumption allowances.
26 The fourth and fifth editions of McCance and Widdowson are available in this format.
27 Caird (1852). Although family incomes in Eden's survey agreed with Caird's pattern, Davies's families showed little variation in family incomes. In both surveys, area F (South-East) was one of the highest income areas. The mean family income in Davies was 8s. 11d per week; in Eden (from the mid-1790s) 13s. 8d.
28 Cobbett ([1830] 1967). See pp. 126, 337 for his views on labourers' diets; pp. 448–9 for his own preferences.
29 Davies (1795: 23).
30 Davies (1795: 21).
31 Davies (1795: 36).
32 Recommended nutrient intakes for calcium exceed 0.5 g. See Department of Health and Social Security (1969: Table 1).
33 The mean may be distorted by the one family in the South-West consuming over 14 lbs/head/week, an amount likely to strain both credence and digestion.
34 Eden (1797: vol. 1, p. 14).
35 Eden (1797: vol. 1, p. 812).
36 Cobbett (1822: 82).
37 Maud F. Davies (1909: 214, 216).
38 *Sixth Report of the Medical Officer of the Privy Council* (1864), 249.
39 Oliver (1895).
40 Dr G. Buchanan's 'Report on the Health of the Distressed Operatives' noted one case of a death from scurvy, see *Fifth Report of the Medical Officer of the Privy Council* (1863: 309). For examples of the diets, see Oddy (1983).
41 See Pember Reeves (1913: 68, 70–1, 143, 156); Rowntree (1901: 135, n. 1).
42 Nitrogen (g) $\times$ 6.25 = protein (g).
43 Oliver (1895: 1635).
44 Merttens (1893–4).
45 See *Report of the Royal Commission on the Supply of Food and Raw Materials in Time of War* (1905).
46 Miller and Payne (1969). A brief account of early experiments with low protein diets between 1887 and 1901 is in Chittenden (1904: 4–7).
47 Miller and Payne (1969: 226).
48 Chittenden (1907: 272).
49 Chittenden (1904: 472–3).
50 *British Medical Journal*, 1906, I, 400.
51 Paul and Greaves (1969).
52 With the exception that in Glasgow, in 1913, Miss Lindsay did give heights and weights of schoolchildren and noted families in which rickety children were to be found (see Lindsay 1913).
53 Department of Health and Social Security (1969: para. 1).
54 Miller and Payne (1969: 227).
55 John Boyd Orr planned the Carnegie Survey of 1937 to have both diet and clinical investigations. See Rowett Research Institute (1955).

References

British Medical Association Committee on Nutrition (1950) *Report of the Committee on Nutrition*, London: British Medical Association.

Burnett, J. (1966) *Plenty and Want: a social history of diet from 1815 to the present day*, London: Nelson.

Caird, J. (1852) *English Agriculture in 1850–51*, London: Longman, Brown, Green & Longmans.

Chittenden, R. H. (1905) *Physiological Economy in Nutrition*, London: Heinemann.

Chittenden, R. H. (1907) *The Nutrition of Man*, London: Heinemann.

Cobbett, W. (1822) *Cottage Economy*, London: C. Clement.

Cobbett, W. ([1830] 1967) *Rural Rides*, Harmondsworth: Penguin.

Davies, D. (1795) *The Case of Labourers in Husbandry*, Bath and London: C. G. & J. Robinson.

Davies, M. F. (1909) *Life in an English Village: an economic and historical survey of the parish of Corsley in Wiltshire*, London: T. F. Unwin.

Department of Health and Social Security (1969) *Recommended Intakes of Nutrients for the United Kingdom*, London: HMSO.

Drake, M. (1969) *Population in Industrialization*, London: Methuen.

Drummond, J. C. and Wilbraham, A. (1957) *The Englishman's Food*, revised edition edited with a new chapter by Dorothy F. Hollingsworth, London: Jonathan Cape.

Eden, F. M. (1797) *The State of the Poor*, 3 vols, London: B. & J. White.

Fifth Report of the Medical Officer of the Privy Council (1863) Appendix V, 1, Parliamentary Papers, XXV, London.

Fussell, G. (1927) 'The change in farm labourers' diet during two centuries', *Economic History* (supplement to *The Economic Journal*), May 1927, 268–74.

Laslett, P. (1965) *The World We Have Lost*, London: Methuen.

Laslett, P. (1971) *The World We Have Lost*, 2nd edn, London: Methuen.

Lindsay, D. E. (1913) *Report upon a Study of the Diet of the Labouring Classes in the City of Glasgow*, Glasgow: Physiological Department, University of Glasgow.

McCance, R. A. and Widdowson, E. M. (1960) *The Composition of Foods*, 3rd revised edn, London: HMSO.

Marshall, D. (1926) *The English Poor in the Eighteenth Century*, London: G. Routledge & Sons.

Merttens, F. (1893–4) 'The hours and the cost of labour in the cotton industry at home and abroad', *Transactions of the Manchester Statistical Society*, 1893–4, pp. 125–79.

Miller, D. S. and Payne, P. R. (1969) 'Assessment of protein requirements by nitrogen balance', *Proceedings of the Nutrition Society*, 28, 232.

Oddy, D. J. (1983) 'Urban famine in nineteenth-century Britain: the effect of the Lancashire Cotton Famine on working-class diet and health,' *Economic History Review*, 2nd series, 36, 68–86.

Oliver, T. (1895) 'The diet of toil', *Lancet*, 1, 1629–35.

Paul, A. and Greaves, J. P. (1969) 'The nutritive value of Professor Chittenden's low-protein diet', *Proceedings of the Nutrition Society*, 28, 15A.

Pember Reeves, M. (1913) *Round about a Pound a Week*, London: G. Bell & Sons.

Report of the Royal Commission on the Supply of Food and Raw Materials in Time of War (1905) Parliamentary Papers 1905 (Cd. 2643) XXXIX, London.

Rowett Research Institute (1955) *Family Diet and Health in Pre-War Britain*, Dunfermline: Carnegie United Kingdom Trust.

Rowntree, B. S. (1901) *Poverty: a study of town life*, London: Macmillan.

Shammas, C. (1990) *The Pre-industrial Consumer in England and America*, Oxford: Clarendon.

Sixth Report of the Medical Officer of the Privy Council (1864) Parliamentary Papers (C. 3416) XXVIII, London.

Watt, J., Freeman, E. J. and Bynum, W. F. (eds) (1981) *Starving Sailors*, London: National Maritime Museum.

Williams, D. E. (1976) 'Were "hunger" rioters really hungry?', *Past and Present*, 71, 70–5.

Woodforde, J. (1978) *The Diary of a Country Parson 1758–1802*, selected and edited by John Beresford, Oxford: Oxford University Press.

7 The achievements of anthropometric history

Roderick Floud

Introduction

Apart from one distinguished but isolated study,[1] anthropometric history began only in the early 1980s, a youngster among academic disciplines. Anthropometry itself has a long history stretching back to the early nineteenth century,[2] but it is only recently that economists and historians have applied the insights of contemporary human biologists and their measurements of the human body to the study of past populations.[3]

Any new scholarly method has to prove its worth and so much of the 1980s and 1990s has been devoted to demonstrating, to the satisfaction of modern social scientists, the value of an observation first made in 1829. In that year, L. R. Villermé (1782–1863), who occupied in France much the same position as a pioneer of public health as Edwin Chadwick did in Britain, observed that:

> Human height becomes greater and growth takes place more rapidly, other things being equal, in proportion as the country is richer, comfort more general, houses, clothes and nourishment better and labour, fatigue and privation during infancy and youth less; in other words, the circumstances which accompany poverty delay the age at which complete stature is reached and stunt adult height.[4]

The same point was made equally clearly by Eveleth and Tanner (1976) who, reporting on the findings of the International Biological Programme, observed that in the modern world:

> A child's growth rate reflects, better than any other single index, his state of health and nutrition, and often indeed his psychological situation also. Similarly, the average value of children's heights and weights reflect accurately the state of a nation's public health and the average nutritional status of its citizens, when appropriate allowance is made for differences, if any, in genetic potential. This is especially so in developing and disintegrating countries. Thus a well-designed growth study is a powerful tool with which to monitor the health of a population, or to pinpoint subgroups of a population whose share in economic or social benefits is less than it might be.[5]

These insights were initially applied to the limited purpose of assessing the health of those men, women and children who migrated from Europe to the New World in the eighteenth and nineteenth centuries, but it soon became apparent that the results of research into the height of populations in the past could be applied to the resolution of a central task of the economic and social history of all countries. That task is to measure and if possible explain the changing living standards of a population, particularly as it passes through a period of economic change such as industrialisation.

The use of anthropometric data in the study of changing living standards in the past certainly did not go unchallenged. Some doubts were based on misunderstandings of the possible roles of genetic and environmental change in changes in human stature occurring over short time periods, but the most serious difficulties arose in reconciling different concepts of the term 'standard of living'. The debate has been a complex one, but it can be characterised (or perhaps caricatured) as a debate between economists and historians. To at least some economists, living standards are synonymous with real wages; that is, they reflect the monetised value of the income of individuals in terms of the goods and services which that individual can buy. Most historians, on the other hand, reject this definition as too narrow; they prefer instead to argue, in biblical terminology, that 'Man does not live by bread alone'. In other words, they prefer to adopt a wider definition of 'standard of living' which includes, for example, health, longevity and even happiness.

Matters were further complicated by the fact that human biologists have often used the term 'nutritional status' to describe the physical state of individuals, reflecting the balance between intake of energy in the form particularly of food and warmth and the demands on the human body for growth, work, play and the conquest of disease. Nutritional status, in this sense, is often seen as synonymous with, or at least measured by, physical growth particularly in height and weight. While human biologists have thus always been clear that 'nutritional status' is a net measure, reflecting the balance between inputs and outputs of energy, the term has confused some historians and economists into believing that differences in human growth between individuals and populations are entirely attributable to differences in nutrition and the food supply.[6]

These matters continue to be debated, but by the end of the 1990s there were some signs of resolution, principally through an acceptance by economists that a wider definition of living standards may be appropriate. As Crafts (1999) puts it:

> Both development economists and economic historians have become increasingly concerned to develop measures of living standards that are more comprehensive than real wages or real GDP per head. Partly, this is because attention has increasingly turned to the lives that people lead rather than the incomes that they enjoy and partly because in most circumstances a substantial element of well-being is derived not on the basis of personal command over resources but depends on provision by the state.[7]

Crafts considers four approaches: 'heights, the Human Development Index (HDI), the Quality of Life Index and imputations to GDP growth.' He argues that:

> they do offer something useful by way of supplementing existing income measures and do tend to confirm that real wages and/or real GDP are not the whole story.[8]

Crafts remains sceptical about the use of height data as an index of welfare *per se*, but is prepared to accept them as diagnostic of situations in which economic growth and rising real wages may be 'misleading indicators of changes in living standards.'[9]

In parallel with these methodological discussions, anthropometric historians have sought to expand the range of data available for analysis, covering longer time periods, more countries and regions, more occupational groups, wider age ranges and both genders. Initially, most data were derived from a single type of source, the records of the armed forces. In some cases, for example in Britain and the USA, these sources recorded mainly volunteers, while in others, for example in most other European countries, they were concerned with conscripts. In either case, however, the records were of males, predominantly in young adulthood and mainly from the working classes. Only rarely did they relate to periods earlier than the middle of the eighteenth century, when many military forces became bureaucratised.

Recent investigations in the Americas have made use of collections of skeletal remains,[10] but the majority of studies have continued to utilise written records of the physical examination of individuals. The initial reliance on military records has been supplemented by the use of records of hospitals, prisons, insurance companies, pension administrations and a miscellany of other sources of anthropometric information. It is particularly useful for analytical and comparative purposes that there has been a substantial expansion in geographical and temporal coverage; Table 7.1 shows the range of written records which had been analysed and published by the beginning of 1999.[11]

Table 7.1 The geographical and temporal coverage of studies of human height

Place	Period	Place	Period
American colonies	1755–63	Japan	1868–1940
Argentina	1770–1865	Korea	1910–45
Australia	1860–1955	Netherlands	1800–1940
China	1900–90	Portugal	1890s
France	1750–1990	Spain	1860–1969
Germany	1880–1939	Sweden	1720–1880
Greece	1927–45	UK	1730–1990
Austria-Hungary	1700–1800	USA	1710–1990
Ireland	1800–45	West Indies	1800–1925
Italy	1861–1938		

In many cases, the use of measurements of human height has contributed to increased understanding of the differences between groups within a given national population. Significant variations in height between regions have been found in most countries, although the extent and direction of such variations has often varied over time. In Britain, for example, the Scots were taller than the English in the eighteenth century and Londoners were particularly short; by the end of the twentieth century this pattern had been entirely reversed.[12] In other countries, for example Italy and Spain, the population has grown significantly overall but the ranking of regional differences has largely been maintained; men and women from southern Italy remain shorter than those from the north. A strong correlation of height with social class is universal, but the extent of inequality in height by class varies considerably in different societies and, very broadly, the more equal a society in terms of income distribution, the taller its population.

The pattern of change in height

While all these findings are of considerable relevance to the economic and social histories of the countries concerned, the findings which have attracted most interest have been those concerned with the overall pattern of growth in heights. Two features in particular deserve discussion.

The first feature is that average height has increased substantially in every country for which evidence is available. It has long been known that sustained economic growth has been accompanied by improvements in literacy and in life expectancy, but it is quite clear that another universal accompaniment is a significant change in the shape of the human body. Table 7.2 shows the rise in average height in a range of countries between 1800 and 1950, since when height has continued to increase, particularly in Japan.

Studies of changing human weight and body mass are still rare, for reasons which will be discussed later, but where they exist and can be compared with height data they show corresponding increases in weight and hence in body mass.[13] Costa and Steckel (1997) suggest that the average BMI of men in their fifties increased in the USA from 23 in 1900 to nearly 27 in 1991,[14] while Floud

Table 7.2 Increase in male height since 1800

Country	Male height in 1800 (cm)	Male height in 1950 (cm)	Change (cm)	Change (%)
UK	168.9	174.1	5.2	3.1
USA	172.9	177.1	4.2	2.4
France	163.7	172.3	8.6	5.3
Netherlands	167.8	178.1	10.3	6.1
Sweden	167.0	177.9	10.9	6.5
Japan	157.1	162.0	4.9	3.1

Source: Steckel and Floud (1997: 424)

(1998) shows a similar increase in Britain.[15] Men and women in Britain in the mid- to late nineteenth century were probably 10–15 per cent lighter than they are today.

While the similarity in trends in height, and probably in weight and body mass, across a wide range of developing countries is striking, it has sometimes been objected that the actual percentage and absolute increases, such as those shown in Table 7.2, are small; what, it has been asked, is all the fuss about? There are two answers to this question.

The first answer is that human beings are extremely good at gauging what, when presented as in Table 7.2, seem to be small differences in height and weight, particularly between themselves and other individuals or groups of people. For this reason, height is one of the invariable components of any police description of a suspected criminal. When we observe a different group of people to ourselves, as for example during foreign travel or in film or television pictures, the difference in appearance produced by systematic differences in height and weight are very striking. It is therefore illuminating to point out that British males born between 1840 and 1859 attained an adult height of 171 cm and an adult weight of 68 kg; this mean height is that attained today by the Bulgarian, Hungarian, Romanian and Russian (Moscow) male populations, but all but one of those populations, the Romanian, is heavier by up to 4 kg than the mid-nineteenth-century British mean.[16] The closest analogue in the modern world to the appearance of a British man of the mid-nineteenth century is a modern Romanian male, who clearly appears both stunted and wasted compared to the modern British male.[17]

The second answer to the question 'What is all the fuss about?' is that there is ample evidence that average height, weight and body mass are correlated with morbidity and mortality, both in modern and in historical populations. Costa and Steckel (1997) summarise the available evidence with respect to the USA:[18]

> The implications of changes in height for mortality have been studied only recently. Costa (1993) found that the functional relationship between height and subsequent mortality is similar among a sample of 322 Union Army recruits measured at ages 23–49 who lived to age 55 and were observed over a 20-year period and among modern Norwegians aged 40–59 observed over a 7-year period.[19] Both the Norwegian curve and the U.S. curve show that mortality first declines with height to reach a minimum at heights close to 185 cm and then starts to rise. . . . A similar relationship is found between height and self-reported health status (Fogel, Costa and Kim 1994). Height appears to be inversely related to height and respiratory diseases and positively related to hormonal cancers (Barker 1992). The relationship between height and mortality remains unchanged controlling for socio-economic covariates, such as occupation, nativity and urbanization. The Norwegian height curve suggests that had the distribution of height in the Union Army sample been the same as in the 1991 National Health Interview Survey (NHIS), older age mortality rates would have fallen by 9 per cent.

. . . The body mass index (BMI) . . . may be an even stronger predictor of productivity, morbidity and mortality than height. The relation between weight and mortality among Union Army veterans measured at ages 50–64 and observed from age 50 until 75 resembles that seen among modern Norwegian males (Costa 1993). Mortality risk first declines rapidly at low weights as BMI increases, stays relatively flat over BMI levels from the low to high 20s, then starts to rise again, but less steeply than at very low BMIs. . . . Among modern American males aged 50–64 the relationship between BMI and self-reported health status, number of bed-days, number of doctors' visits and number of hospitalizations follows a similar U-shaped pattern (Costa 1996). Costa (1993) argued that the low weights of Union Army veterans can partially explain why mortality for their cohorts was higher than for cohorts today. Had it been possible to shift the BMI distribution of Union Army veterans one standard deviation to the right so that the mean would be equivalent to that prevailing in modern Norway, the implied 14 per cent reduction in the mortality rate would explain roughly 20 per cent of the total decline in mortality above age 50 from 1900 to 1986, a percentage greater than that explained by changes in height.[20]

In other words, changes in height and weight, of the extent observed in historical populations, matter because they are related to so many other attributes and capabilities of human beings. The changes in heights and weights which have occurred over the last three centuries thus help to explain many other aspects of changes to the human condition. There is little doubt, for example, that the continuing rise in recent decades in the life expectancy of the British population is related to improvements in the nutritional status of the population during its childhood some decades ago.[21]

The emerging picture of changes in height and weight over several centuries thus has great intrinsic interest. But the second general finding of two decades of research in anthropometric history is that growth in height, weight and BMI has not been uniform. The speed of growth has varied and there have indeed been episodes in several countries during which the average height of a population has fallen. Steckel and Floud (1997) list relevant data which show that height declined during periods of industrialisation, in particular in Britain between the 1820s and the 1850s, in the USA between 1830 and 1890 and in Australia between 1867 and 1893; there may also have been a short-lived decline in height in Germany between the 1860s and the 1880s. On the other hand, no similar declines in average stature seem to have occurred in France, the Netherlands or Sweden. Irrespective of whether height actually declined, the speed of increase in height has varied in all countries; there has been no uniform 'secular trend'. As a consequence, a 'league table' of countries arrayed according to their average height shows substantial changes over time, with the Netherlands and Sweden, for example, coming from behind to overtake the early leaders, Britain and the USA.[22]

Height and the 'standard of living' debate

These findings, and particularly those concerned with declines in average height in the nineteenth century in Britain and the USA, have provoked considerable debate. In large measure, this was because of the historiographical context, in particular that within British economic history. As Stanley Engerman (1997) puts it:

> Probably the most famous debate on economic change has been that known as 'the standard of living debate,' about the impact of the British industrial revolution. This debate began among individuals living in those times, continues today and will go on, no doubt, tomorrow. The debate's vehemence has several distinct sources – the politics of the British class struggle, comparisons of British growth with the pattern of communist economic growth in the twentieth century, and more general questions of the attitudes toward society and culture that emerged in modern times.[23]

By the late 1980s, this debate, which had begun with the writings of Marx and Engels nearly 150 years before and had continued to divide British economic historians into 'optimists' and 'pessimists', had reached an unstable and uneasy consensus; in very general terms, it was accepted that the living standards of the British working class had hardly benefited from the early years of industrialisation. On the other hand, it was thought that, from the 1830s or 1840s, real wages had begun to improve and thereafter continued to rise throughout the remainder of the nineteenth century.

It was this consensus which was challenged by the publication in 1990 of *Height, Health and History: Nutritional Status in the United Kingdom, 1750–1980*,[24] and specifically by its finding that there was 'a decline in mean heights of those born between the 1820s and 1850s' and that this 'is a strong argument indeed for the pessimistic interpretation of the impact of industrialisation on the British working class.'[25] This statement attracted not only considerable attention but also scepticism, based partly on the methodological grounds discussed above – that the proper means of measuring the impact of industrialisation was through the calculation of real wages – but also because it was, and remains, difficult to identify the precise cause of the decline in heights which was observed.

This difficulty is endemic to anthropometric history. Just because the nutritional status of an individual or of a group is the product of so many environmental and genetic factors acting upon the human body, very few or any of which remain constant for more than a short time, it is normally impossible to assign a single cause to an observed change in mean height or weight. Thus it was impossible to answer, simply or unequivocally, the question: 'Why did average heights fall in Britain between 1820 and 1850?' and the failure so to do threw doubt on the utility of the method. The same difficulty, and thus the same scepticism, was experienced in the USA, where the issue became known as the 'antebellum paradox'; why, at a time when real incomes were apparently rising in the years before the American Civil War, should heights apparently have fallen.[26]

However, the fact that the same phenomenon of declining height at a time of apparently increasing real incomes was observed in both the two countries which experienced early industrialisation and early urbanisation gave clues to its likely cause. In the British case, a pointer to possible causes lay in the fact that expectation of life at birth, which had been rising from the end of the eighteenth century, reached a plateau in the 1830s from which it did not rise again until the 1870s. Essentially, it seems likely that economic growth in the form of industrialisation produced, in both Britain and the USA, towns and cities whose sanitary infrastructure was initially weak and which was unable to cope with rapid growth. In an age before the development of the germ theory of disease, it took some time before the pragmatic observations of men like Edwin Chadwick were translated into adequate drainage and sanitation.[27] In the meantime, the towns and cities of mid-nineteenth-century Britain and USA became breeding grounds of endemic and epidemic disease. Mortality rose, or at least ceased to fall, and the conquest of disease became a major call on the bodies of children, inhibiting their growth and leading to declining average adult stature in the population. Thus, although the adult workers in the industrialising towns benefited from higher wages, their children's bodies suffered from the environment in which they grew up.

While, because of the nature of the evidence, much must remain supposition, the overall conclusion has recently gained support from reworking of the evidence of changes in real wages. Careful analysis of the available data led Feinstein (1998) to conclude that:

> Wage earners' average real incomes were broadly stagnant for 50 years until the early 1830s, despite the fact that in many parts of the country they were starting from a very low level, having been falling in the second half of the eighteenth century. Some slight progress was made in the mid-1830s, but earnings then fell back again in the cyclical depression during 1838/42, and it was not until the mid-1840s that they at last started an ascent to a new height. More substantial gains were not achieved until the 1860s, and it was only after the post-1873 downturn in prices that average real earnings finally accelerated.
>
> The picture looks even more bleak when we take account of a number of other features of the period which would have had an adverse effect on living standards.[28]

These factors were adulteration of food and drink, the increase in the number of dependents supported by each adult worker, poor housing, inadequate public health and the decline in relief expenditure under the Poor Law. Feinstein concludes:

> The combined effect of these three factors would thus reduce any improvement in the standard of living of the average working-class family in the United Kingdom between the 1780s and the 1850s from about 30 per cent to somewhere in the range of 10 to 15 per cent.[29]

Feinstein's judgement is that this account of changes in living standards

> eliminates the paradox of the decline in nutritional status indicated by the early nineteenth century height data occurring at a time of an allegedly swift advance in living standards. . . . For the majority of the working class the historical reality was that they had to endure almost a century of hard toil with little or no advance from a low base before they really began to share in any of the benefits of the economic transformation they had helped to create.[30]

The data for the calculation of real wages are even more problematic in the USA than in Britain, but it seems likely that many of the same factors affected the standard of living of the urban population and thus contributed to the 'antebellum paradox' of declining heights. Certainly the US cities experienced high rates of epidemic disease before public health measures took effect towards the end of the nineteenth century. Although it is unlikely that certainty will ever be achieved, the evidence of declining heights has thus already made a significant contribution to one of the most enduring issues of modern economic history.

Future challenges

Despite these achievements of anthropometric history in its first twenty years, much remains to be done. In part, current deficiencies in our knowledge stem from the defects of source materials. Although scholars have been increasingly ingenious in seeking out anthropometric data, it remains the case that military records of various kinds provide the main source. Such records almost always exclude women and relate to men in young adulthood; only rarely do these sources provide information on adolescents, which might allow delineation of the adolescent growth spurt, or on older men. Military records, again, often relate only to the working-class population, since the middle and upper classes were either able to avoid military service or served as officers, who were often not measured.

A concomitant difficulty is that few of the sources record information on anthropometric features other than height. This is not true of one of the most informative of sources, the records of the Union Army in the American Civil War, which describes the medical examination of many characteristics of thousands of veterans.[31] But few of the data sources which have hitherto been published or analysed, for any country, record even the weight of those who were measured, perhaps because weight was of less use as a form of identification or because it varied less between young males.

In some cases, however, it is possible to find and use data on weight and the derivative of height and weight, body mass. Costa and Steckel (1997) report evidence on weight and body mass from the USA, principally from the records of the Union Army, while Floud (1998) gathered numerous observations of British heights and weights from secondary sources. These latter data suggest that men and women of the late nineteenth century in Britain, by comparison with modern standards, were even more wasted than stunted. Although heights began to rise in the late

nineteenth century, after the decline discussed above, it was not until after the beginning of the twentieth century that weights followed, allowing body mass to move
towards modern levels. Since Waaler's (1984) research in Norway and the initial
results of the analysis of the Union Army both suggest that height and body mass
have independent effects on morbidity and mortality, the collection of more data
on historical weights should have a high priority in future anthropometric research.

A more important task, however, is to develop our understanding of the complex interrelations between anthropometric data and other variables of interest
to economists and historians.[32] The nutritional status of individuals and of groups
within a population is determined, at a first approximation, by the energy resources
– principally food and warmth – available to those individuals and groups and
the interaction of those energy resources with the demands of body maintenance,
work, play and the conquest of disease. Thus an overall constraint on nutritional
status is provided at any one time by the output of agriculture and of fuel supplied
at prices which the population can afford. But, particularly in underdeveloped
societies which are highly dependent on agricultural goods produced by labour
power, the energy which can be devoted to agricultural work is itself dependent
on the nutritional status and bodily strength of the members of the population.
There is thus a circularity in the relationship between nutritional status and
energy supply. But the position is even more complex because human societies
typically do not distribute goods equally across the whole population; command
over resources of income and wealth is unequal and, in most societies in the
past, significant proportions of the population had such meagre resources that
they were stunted and wasted, unable to work for more than a few hours a day.
In such societies, the circularity between nutritional status and energy supply can
be a vicious one, in which the lack of command over resources of most of the
population leads to low productivity, low output and a stagnant economy.

The process of economic development can be seen as one of escape from such
a vicious circle. If agricultural output rises, for whatever reason, and if the benefits of this are diffused rather than captured by a small segment of the population,
then it may be possible to enter a virtuous circle. A growing food supply permits
an improvement in nutritional status, which then itself permits individuals to
work harder to produce yet more output. Possession of even a small surplus over
basic food needs allows for increasing diversification of production into other forms
of output, such as manufacture and the provision of services, which themselves
benefit from a workforce with improved nutritional status, greater strength and
health, longer life and perhaps even greater intelligence.

Seen in this light, the nutritional status of individuals and groups has a crucial
status within the explanation of modern economic growth. But it is much easier
to assert such a status than to demonstrate it in detail in studies of particular
societies and time periods, particularly given the paucity of data on some of the
significant variables. Anthropometric history has contributed much in the 1980s
and 1990s to the measurement of changing living standards. Its task, in the next
twenty years, will be to supply some of the evidence and the insights which will
allow the complexity of economic growth in the past to be unravelled.

Notes

1 Le Roy Ladurie (1973).
2 Tanner (1981).
3 Three compendia of articles in the field of anthropometric history are Komlos (1995), Komlos and Baten (1998) and Komlos and Cuff (1998).
4 Villermé (1829: 385), trans. Tanner (1981: 162).
5 Eveleth and Tanner (1976: 1).
6 Matters were further confused by the invention by some anthropometric historians of the term 'the biological standard of living' as a synonym for nutritional status (see Komlos 1995).
7 Crafts (1999: 11).
8 Crafts (1999: 12).
9 Crafts (1999: 14).
10 Steckel *et al.* (1998).
11 An earlier discussion of the implications of many of these data sets can be found in Steckel and Floud (1997).
12 Floud *et al.* (1990).
13 Costa and Steckel (1997: 53–7); Floud (1998).
14 Costa and Steckel (1997: 55).
15 Floud (1998: Figure 14).
16 Eveleth and Tanner (1976: 1).
17 The terms 'stunted' and 'wasted' are used in their technical sense of being significantly below mean height and weight.
18 Costa and Steckel (1997: 52–3). References have been changed to correspond with those in the bibliography.
19 The Norwegian population is used for comparison because this population provides the largest available data set, that reported by Waaler (1984).
20 When height and weight are jointly related to subsequent mortality, the predicted decline explains 15 per cent of the total decline in mortality. When height and weight are jointly related to subsequent morbidity, the predicted decline explains 35 per cent of the total decline in morbidity (Fogel *et al.* 1994).
21 Institute of Actuaries (1999).
22 Steckel and Floud (1997: 430, 424).
23 Engerman (1997: 17); further summaries and discussions of the debate can be found in Lindert (1994) and Engerman (1994).
24 Floud *et al.* (1990).
25 Floud *et al.* (1990: 305).
26 Recent contributions to the debate on the antebellum paradox include Haines (1998), Craig and Weiss (1998) and A'Hearn (1998).
27 Halliday (1999).
28 Feinstein (1998: 649).
29 Feinstein (1998: 650).
30 Feinstein (1998: 652).
31 Center for Population Economics (1999).
32 Fogel *et al.* (1996).

References

A'Hearn, B. (1998) 'The antebellum puzzle revisited: a new look at the physical stature of Union Army recruits during the Civil War', in Komlos, J. and Baten, J. (eds) *The Biological Standard of Living in Comparative Perspective*, Stuttgart: Franz Steiner Verlag, pp. 250–66.

Barker, D. J. P. (ed.) (1992) *Fetal and Infant Origins of Adult Disease*, London: British Medical Journal.

Center for Population Economics (1999) *Public Use Tape on the Aging of Veterans of the Union Army*, Chicago: Center for Population Economics, University of Chicago.

Costa, D. L. (1993) 'Height, weight, wartime stress and older age mortality: evidence from the Union Army records', *Explorations in Economic History*, 30, 424–49.

Costa, D. L. (1996) 'Health and labour force participation of older men, 1900–1991', *Journal of Economic History*, 56, 62–89.

Costa, D. L. and Steckel, R. H. (1997) 'Long-term trends in health, welfare and economic growth in the United States', in Steckel, R. H. and Floud, R. C. (eds) *Health and Welfare during Industrialization*, Chicago: University of Chicago Press, pp. 47–91.

Crafts, N. F. R. (1999) 'Quantitative economic history', *L. S. E. Working Papers in Economic History* 48/99, London: London School of Economics.

Craig, L. A. and Weiss, T. (1998) 'Nutritional status and agricultural surpluses in the antebellum United States', in Komlos, J. and Baten, J. (eds) *The Biological Standard of Living in Comparative Perspective*, Stuttgart: Franz Steiner Verlag, pp. 190–207.

Engerman, S. L. (1994) 'Reflections on "the standard of living debate": new arguments and new evidence', in James, J. A. and Thomas, M. (eds) *Capitalism in Context: essays on economic development and cultural change in honor of R. M. Hartwell*, Chicago: University of Chicago Press, pp. 50–79.

Engerman, S. L. (1997) 'The standard of living debate in international perspective: measures and indicators', in Steckel, R. H. and Floud, R. C. (eds) *Health and Welfare during Industrialization*, Chicago: University of Chicago Press, pp. 17–45.

Eveleth, P. B. and Tanner, J. M. (1976) *Worldwide Variations in Human Growth*, Cambridge: Cambridge University Press.

Feinstein, C. H. (1998) 'Pessimism perpetuated: real wages and the standard of living in Britain during and after the Industrial Revolution', *Journal of Economic History*, 58, 625–59.

Floud, R. C. (1998) *Height, Weight and Body Mass of the British Population since 1820*, Working Papers on Historical Factors in Long Run Growth no. 108, National Bureau of Economic Research.

Floud, R. C. and McCloskey, D. N. (eds) (1994) *The Economic History of Britain since 1700*, 2nd edn, Cambridge: Cambridge University Press.

Floud, R. C., Wachter, K. and Gregory, A. (1990) *Height, Health and History: nutritional status in the United Kingdom, 1750–1980*, Cambridge: Cambridge University Press.

Fogel, R. W., Costa, D. L. and Kim, J. M. (1994) 'Secular trends in the distribution of chronic conditions and disabilities at young adult and late ages, 1860–1988: some preliminary findings', unpublished ms, University of Chicago.

Fogel, R. W., Floud, R. C., Costa, D. L., Kim, J. M. and Acito, C. J. (1996) 'A theory of multiple equilibria between populations and food supplies: nutrition, mortality and economic growth in France, Britain and the United States, 1700–1980', unpublished ms, University of Chicago.

Haines, M. R. (1998) 'Health, height, nutrition and mortality: evidence on the "antebellum puzzle" from Union Army recruits for New York State and the United States', in Komlos, J. and Baten, J. (eds) *The Biological Standard of Living in Comparative Perspective*, Stuttgart: Franz Steiner Verlag, pp. 155–80.

Halliday, S. (1999) *The Great Stink of London*, Stroud: Sutton.

Komlos, J. (ed.) (1995) *The Biological Standard of Living on Three Continents: further explorations in anthropometric history*, Boulder, CO: Westview Press.

Komlos, J. and Baten, J. (eds) (1998) *The Biological Standard of Living in Comparative Perspective*, Stuttgart: Franz Steiner Verlag.

Komlos, J. and Cuff, T. (eds) (1998) *Classics in Anthropometric History*, St Katharinen: Scripta Mercaturae.

Institute of Actuaries (1999) *Report of the Continuous Mortality Investigation Bureau*, London: Institute of Actuaries.

Le Roy Ladurie, E. (1973) 'Anthropologie de la jeunesse masculine en France au niveau d'une cartographie cantonale (1819–30)', in *Le Territoire de l'Historien*, Paris: Gallimard.

Lindert, P. (1994) 'Unequal living standards', in Floud, R. C. and McCloskey, D. N. (eds) *The Economic History of Britain since 1700*, 2nd edn, Cambridge: Cambridge University Press, vol. 1, pp. 357–86.

Steckel, R. H. and Floud, R. C. (1997) 'Conclusions', in Steckel, R. H. and Floud, R. C. (eds) *Health and Welfare during Industrialization*, Chicago: University of Chicago Press, pp. 423–49.

Steckel, R. H., Sciulli, P. W. and Rose, J. C. (1998) 'Skeletal remains, health and history: a project on long term trends in the western hemisphere', in Komlos, J. and Baten, J. (eds) *The Biological Standard of Living in Comparative Perspective*, Stuttgart: Franz Steiner Verlag, pp. 139–54.

Tanner, J. M. (1981) *A History of the Study of Human Growth*, Cambridge: Cambridge University Press.

Villermé, L. R. (1829) 'Mémoire sur la taille de l'homme en France', *Annales d'hygiène publique*, 1, 551–99.

Waaler, H. T. (1984) 'Height, weight and mortality: the Norwegian experience', *Acta Medica Scandinavica*, 679, supplement, pp. 1–56.

8 Palaeopathology as a tool for the study of survival of past populations

Pia Bennike

Introduction

It may seem paradoxical to address the issue of survival in past populations exclusively on the basis of skeletal remains of individuals who are no longer alive. However, survival not only deals with death and the calculation of life expectancy derived from the distribution of skeletal remains, but also deals with surviving any number of risks and traumas experienced during life, such as acts of violence, serious accidents or various diseases. They may all leave traces on the skeletons or teeth thereby reflecting the dangers of life in a particular population. Such skeletal traces are a consistent part of palaeopathological studies. Both lesions as seen on bones and teeth, and demographic patterns, including longevity and fertility of past populations, must be included in order to elucidate the issue of survival. Furthermore, a study of survival in human populations cannot be addressed only from a strictly biological angle, as the influence of various cultural aspects must also be taken into consideration.

This means that both the biological and cultural levels at which we want to study the survival of a certain population must be defined. The biological level might be our own genus, species, subspecies, population, or it may be limited to a family or an individual. Each level has its own specific aspects. For example, a study of survival at the family level must take into account whether all members of the family are closely related to each other genetically or not. Furthermore, the social organisation within the studied population must be considered. When dealing with a study of survival, one may also run into 'clashes' such as a large number of people may survive thanks to the death of a number of warriors who defended them in battle, a mother may die during childbirth while her baby survives as a member of the next generation, and an individual who dies of a severe, infectious disease may eliminate any further contamination, whereby the rest of the group to which he or she belongs survives.

Instead of describing the theoretical issues of survival which are rather complex, the discussion will be limited to six different populations from three separate periods. One of the two populations from each period is considered as consisting of non-survivors and the other of survivors. Each period, however, deals with populations at different levels; the first period deals with species, the second with population groups belonging to the Mesolithic and Neolithic cultural complexes, and the third period deals with population groups with diverse racial backgrounds, the caucasoid or mongoloid races.

Neanderthals and the Upper Palaeolithic populations

The two populations from the earliest period, c.35,000–30,000 years ago, are the Neanderthals and the Upper Palaeolithic people. The latter belong to the early anatomically modern humans in Europe and are supposed to be the direct ancestors of present day European populations. With regard to survival, these two populations are well suited for comparison because of their very different destinies.

There is still some discussion as to whether the Neanderthals were totally replaced by the more successful immigrants or whether they were more or less absorbed into the immigrating stock (Wolpoff and Caspari 1997; Wolpoff 1998). However, most theories and the majority of scientists agree that the Neanderthals became extinct (Klein 1995; Lahr 1996). The fact that neither we nor our ancestors seem to have inherited any specific Neanderthal genes has been suggested in a recent study based on analysis of mitochondrial DNA (Krings *et al.* 1997). The two populations, both belonging to the genus *Homo* and probably to the same species *sapiens*, therefore seem to have been genetically divergent at a certain level of the category sub-species. The theories on why the Neanderthals became extinct and the possible advantages substained by the *Homo sapiens sapiens* are numerous. One interesting theory put forward by Steven Mithen (1996) and several others is that a substantial structural change in the brain permitted a more abstract and qualified communication and conscience. This probably improved the pattern of social organisation and behaviour of the anatomically modern humans. It also allowed a revolutionary combination of knowledge into abstract forms and symbols which the cultural remains so clearly illustrate (such as an animal as a religious spirit, a figurine as a symbol, a carved piece of bone or stone as an indication of social status). The intellectual advantages of *Homo sapiens sapiens* probably left the Neanderthals with little chance of survival.

Palaeopathology and palaeodemography of Neanderthals

In 1995, Eric Trinkaus published a study on mortality distributions of 206 European and Near Eastern Neanderthals. Associated bones from a total of 40 skeletons and 166 isolated elements were compared with those of 11 recent human samples. The results indicate a bias in the Neanderthal sample, which had too few infants and older adults plus too many adolescents and very young adults. Whereas the immature mortality may produce a distribution within the ranges of more recent skeletal populations, Trinkaus (1995) suggested that the high mortality of the very young adults and the few remaining older adults is the product of a combination of demographic stress, which may be associated with the high levels of stress indicators found in Neanderthal skeletons, and the need for full mobility among all individuals and hence the dearth of older people dying in shelters. This interpretation of palaeodemographic data is consistent with the evidence of high levels of adaptive stress which have previously been suggested by palaeopathological analyses of Neanderthal remains (Trinkaus and Zimmermann 1982). Even though a few remains of old individuals have been found, like the early find from Capelle aux Saints in France, very few were more than 40 years old at the time of death (Trinkaus 1985).

A study of traumatic lesions carried out on bones from 17 Neanderthal individuals by Berger and Trinkaus (1993) revealed a total of 27 injuries. These were mainly distributed on the head and upper body with fewer on the lower limbs. This pattern of injuries was compared with the pattern suffered by a variety of active modern human groups. Only rodeo riders, who incur the most injuries among modern humans, fitted the Neanderthal pattern statistically. As the Neanderthals had heavier and more robust bones, they must have been involved in acts of considerable violence. No doubt the Neanderthals ran significant risks while hunting large animals such as bison and elk, because they had to get very close to their prey in order to thrust their heavy spears (Frayer 1981).

According to several studies, the first Neanderthal skeletal remains which were found in the Neander Valley were described as belonging to a battered man; the left arm had been broken to the extent of malfunction and several joints showed changes due to severe arthritis.

Neanderthals can mainly be distinguished from all other populations, fossil or living by the typical morphology of the skull and the face. The shape and wear of the front teeth indicate that the front of the mouth functioned routinely as a clamp or vice, and some studies suggest that the specific structure and position of the face may have been related to these functions (Rak 1987; Demes 1987).

Like any early people, the Neanderthals had a robust postcranial skeleton, suggesting both extraordinary muscular strength and remarkable endurance. Their forearms and lower legs were relatively short, perhaps as a functional correlate of general limb robusticity or perhaps as part of a physical adaptation to the generally cool climate in which they evolved (Churchill 1998; Hoffecker 1999). The bowed femoral bones were formerly interpreted incorrectly as a result of rickets due to vitamin-D deficiency. The shape is more likely related to muscle strength and the weight of the skeleton. The remarkable cortical thickness compared to that of modern femurs seem to confirm this (Ruff 1992).

In terms of behaviour, some scientists have found that, contrary to the theory that old people were left to die in shelters, the Neanderthals were strikingly modern in their apparent care for individuals who were handicapped or incapacitated long before death. A number of skeletons with severe pathological changes, such as those found in Shanidar in Iraq, seem to reflect that the individuals were cared for (Trinkaus and Zimmermann 1982; Crubézy and Trinkaus 1992). One upper arm with evidence of long-term immobility and probable paralysis of the upper limb is a good example. However in most detectable aspects, the Neanderthals were probably still very primitive. Their graves were very simple, and some think that there is a lack of any clear evidence of ritual or ceremony. No religious or ideological motivation need to have been implied, and from a less romantic point of view, burials may have simply served to remove bodies from sight (Klein 1989). Furthermore, the remains of flowers revealed by pollen may have landed in the graves by chance.

Neanderthal populations were relatively small in number. They were probably unable to colonize the harshest, most continental parts of Eurasia for reasons like not being able to construct 'permanent' well-heated dwellings (Klein 1989). In this and most other archaeologically observable aspects, they remained similar to their predecessors and very unlike their anatomically modern successors, who may even have rejected them as partners. However, it is still not known whether the two groups were members of the same subspecies *sapiens* and able to produce fertile progeny.

The question of what happened to the Neanderthals and the closely related questions of where and when modern people originated still remain partly unresolved. Apparently anatomically modern people had behavioural capacities which differed from those of the Neanderthals. Eventually, the previously mentioned neurological change which unfortunately is not detectable in the fossil record, enabled modern people to develop a capacity for culture and social organisation which gave them a clear adaptive advantage over the Neanderthals and all other non-modern people. The result was that they spread throughout the world, physically replacing all non-moderns of whom the last to succumb were perhaps the Neanderthals of western Europe. The change in climatic conditions and the subsequent new subsistence economy together with the arrival of modern people, gave the Neanderthals only a slim chance of survival; and if they had any strategy at all, it probably failed.

The Upper Palaeolithic population

The extensive cultural variability of the past 30,000 years almost certainly required the existence of modern people, with their seemingly infinite capacity for innovation. The vital statistics of the Upper Palaeolithic people can be only crudely estimated because of the relatively small number of skeletons with established age and gender. However, according to Vallois (1961) the common mortality patterns of the composite Upper Palaeolithic skeletons from various periods and places and the remarkable, but later series of 163 skeletons from Taforalt in Morocco, seem to resemble that of most later prehistoric and historic hunter-gatherers.

Child mortality has been reported to be high. It has been suggested that women apparently had a shorter life expectancy than did men, probably because of the risks associated with childbearing. Many women rarely reached the age of 40 and men probably rarely reached 60 or even 50 years of age; but significantly, maximum life expectancy probably exceeded that of the Neanderthals, perhaps by as much as 20 per cent, as suggested by Klein (1989). As a consequence, Upper Palaeolithic human groups probably contained more older individuals, whose accumulated knowledge could promote group survival, particulary in times of crises.

Unlike Neanderthal skeletons, Upper Palaeolithic ones are rarely described as showing evidence of numerous serious accidents or diseases, suggesting that their culture provided a far more effective shield against environmentally induced trauma. Skeletal anomalies that could reveal the cause of death are particularly rare, but include conspicuous lesions such as those seen on the skull, mandible, pelvis and femur of the famous Old Man skeleton from Cro-Magnon, a man in his forties. According to Dastugue (1982) the lesions might be attributed to a severe *ante mortem* fungal infection. In other skeletons it has been suggested that dental abscesses may have caused fatal blood infections, but this theory provides little skeletal evidence. In contrast, most other Upper Palaeolithic skeletons rarely reveal such finds, and most of them are reported to have had a relatively healthy dental status, probably because their diets included few foods that encouraged caries or plaque formation.

Reports on Upper Palaeolithic skeletal anomalies include an Italian skeleton of a male adolescent who apparently suffered from dwarfism (*acromesomelic dysplasia*). It has been suggested that this disease may have reduced his life expectancy under Palaeolithic conditions (Frayer *et al.* 1988).

Other known Upper Palaeolithic skeletal abnormalities were debilitating to various degrees but probably not fatal as shown in the following. None indicate epidemic diseases, which were probably rare until the later increase in population densities. The 27,000-year-old Czech examples from Dolní Vestonice include a female mandible with a deformed left temporomandibular joint suggesting partial facial paralysis (Klima 1987). Another woman from the site, who was laid out between two very young men in a triple grave, had a shortened, deformed right femur, probably accompanied by a curvature (scoliosis) of the vertebral column and several other abnormalities (Formicola *et al.* 1998). They are thought to be due to a rare, inherited disorder characterised by disturbed ossifications of the epiphysis during childhood. Several skeletons from Cro-Magnon and two other sites in France exhibit fused cervical vertebrae. A healed fracture, bone lesions and degeneration implying a permanently dislocated left shoulder have been reported by Dastugue and de Lumley (1976). The skeleton from Chancelade has a laterally deviated right big toe (hallux valgus), and a skeleton from Combe-Capelle has an asymmetric sacrum reflecting a lateral curvature of the spine (scoliosis) (Dastugue and de Lumley 1976). By modern analogy, the occurrence of cervical fusion may indicate that some older Upper Palaeolithic individuals had a relatively sedentary life or literally remained seated most of the time. Similarly, the laterally deviated toe may reflect poorly fitting footwear (Klein 1989), although in my opinion there may be other explanations as well.

Roper (1969) found that skeletal evidence for deliberate injury is relatively rare, probably because Upper Palaeolithic hunter-gatherers, like most other ethnographically recorded hunter-gatherers, rarely engaged in warfare or interpersonal violence. In some instances, fractures of the skull could clearly have healed spontaneously, and many other bones may have fractured or been crushed post mortem.

However, although evidence of violence is rare, it does exist. Both the Old Man from Cro-Magnon and the man buried at Chancelade were probably too disabled to fend for themselves. Their survival may suggest that Upper Palaeolithic people, like Neanderthals before them and people from later periods, cared for their old and sick. Such care need not have been entirely philanthropic, since older people in hunter-gatherer societies commonly possess vital knowledge and experience (Klein 1989). However, care in a more emotional, abstract sense, probably not practised by the Neanderthals, may have been needed for the survival of the Dolní Vestonice woman with the deformed temporomandibular joint. It has been suggested that her face may have resembled the one engraved on a fragment of ivory and the face of a sculpted clay head found nearby, which both droop on the left side, just as hers probably did. That the two figures may represent the oldest known human portraits as has been suggested in some popular publications is, however, arguable.

The overall size of the human body decreased during the Upper Palaeolithic period (Frayer 1984). It is thought to be due not only to the development and improvement of technologies, but also to behavioural changes. In other words, it is possible that natural selection favouring the big and strong came to an end.

The Mesolithic and Neolithic populations

In contrast to the Neanderthals, the other five populations considered in this chapter all belong to the modern species *Homo sapiens sapiens*. Populations from the Mesolithic and Neolithic periods have been chosen as examples from the second period *c.*12,000–6000 years ago. Comparing these two peoples is very different from comparing Neanderthals and early European anatomically modern humans. Numerous cultural replacements in many geographical areas may have played a more dominant role than any actual genetic or demic replacement during this period. In most areas of Europe, hunting, and perhaps fishing and gathering of the Mesolithic Age, was followed by the agricultural and cattle breeding subsistence economy of the Neolithic Age. Many aspects, such as the degree of continuity of the Mesolithic population are still being discussed, and many questions are still unanswered. It is not yet clearly understood whether the Mesolithic populations were pushed to marginal areas by immigrating farmers where they decreased in numbers (the demic expansion theory) as supported by Ammermann and Cavalli-Sforza (1984), Barbujani *et al.* (1995), or whether they adopted agriculture and continued more or less as the Neolithic population, in which case the genetic pool would have remained largely unchanged (Harding *et al.* 1990; Richards *et al.* 1996). This issue is pivotal when discussing survival patterns. The first theory implies that the Mesolithic population would be the non-survivors, whereas the second theory implies that they were survivors who simply evolved into the Neolithic population (the cultural exchange theory). It may appear to be a rather simple task to put these theories to the test by examining the cultural findings and perhaps more importantly, the skeletal remains from the two periods. Such studies must be based on relevant biological techniques, including recently developed DNA analyses and statistics. A number of studies of this nature have appeared in the literature, but reveal a wide range of contradictory results which might be explained by factors such as regional variability. To compare survival patterns in the Middle East, where agriculture is believed to have arisen, with those in northern European areas several thousand years later may be almost impossible (Jackes *et al.* 1997). In Denmark, for example, the increasing number of finds of skeletons from the Mesolithic and Neolithic definitely exhibit a number of differences between the two populations. The two main differences are seen in the stature and build. The early Neolithic population is generally smaller and more gracile and the teeth decrease in size. However, the skeletal material is not extensive enough to provide us with answers to whether the differences are based on genetic changes or changes in the subsistence economy and lifestyle. Future DNA analyses may show us to what extent the genes of the Mesolithic populations survived in the various regions.

With few exceptions, the introduction to agriculture in Europe has traditionally been regarded as a replacement of hunter-gatherer Mesolithic cultures by Neolithic farmers. Within this framework, Neolithic cultures have been viewed as the result of colonisation by farmers from the south or south-east. The role of the resident hunter-gatherers has often been disregarded or minimised as a consequence of such views. Yet, colonisation models for the origin of farming in Europe entail assumptions which cannot be verified by archaeological data alone. The opposing view, where indigenous adoption of farming has been contemplated for the transition, usually requires explanations concerned with environmental or demographic stress, or social competition (Straus 1995).

Many assumptions regarding the Mesolithic–Neolithic transition have been examined on archaeological grounds and are often subject to contrary interpretations. The study of human remains within the biocultural framework provides convergent information which is directly related to the biological and ecological conditions of the people involved. Skeletal stress indicators and pathologies may reflect the health status in the Mesolithic and the Neolithic populations. In general, most studies have shown that there are few signs of poor health or dietary stress in Mesolithic populations, and the rates of pathological changes of the European Mesolithic do not seem to be particularly high compared to the Neolithic populations.

The Mesolithic population

In 1984 Constandse-Westermann *et al.* studied pathological changes and stress indicators with regard to the mortality pattern of the Mesolithic population based on 316 skeletal remains. They concluded that the Mesolithic skeletal evidence does not indicate any specific factor which would have contributed substantially to low population growth relative to that of the subsequent Neolithic population.

They found, in agreement with many other studies, that the frequencies of nutritional and infectious diseases in the western European Mesolithic skeletal samples were low when compared to the Neolithic samples. Although Neolithic teeth often presented less attrition than Mesolithic ones, the incidence of dental decay, periodental diseases and tooth loss was higher in the former (Meicklejohn and Zwelebil 1991).

Several instances of osteoarthritis have been reported in both hunterer-gatherer groups and Neolithic groups. However such skeletal changes, like many other pathological changes, do not usually lead to the immediate death of the individuals in question; but several skeletons from both periods provide clear evidence that the individuals were handicapped. They were able to survive until a fairly advanced age with a considerable degree of degenerative disablement. There seems to be a difference between the Mesolithic and Neolithic population with regard to degenerative diseases. In some studies osteoarthritis or osteophytosis seems to have been somewhat more frequent in Mesolithic skeletons (Larsen 1995) but other studies seems to show contradicting results (Constandse-Westermann *et al.* 1984). It has been suggested that a higher bone-mass and pronounced muscle attachments on these prehistoric skeletons is an expression of life-long physical exertion (Ruff *et al.* 1984).

Traumatic injuries were more frequent but in most cases non-lethal in the Mesolithic. The majority of injuries related to interpersonal violence seem to have been inflicted in the later phases of the period when settlements seemingly became larger. Constandse-Westermann *et al.* (1984) suggested that the transition toward a more sedentary life was apparently initiated in the last phases of the Mesolithic and that population densities were approaching those of the Neolithic. The authors concluded that the primary mechanism responsible for the difference between hunter-gatherer populations and agricultural/pastoralist groups, was a decrease in the fertility rate rather than an increased mortality rate. They found no evidence of higher mortality levels in the Mesolithic compared to the Neolithic.

Most evidence of homicides in hunter-gatherer and agricultural groups are randomly directed at single individuals, often a victim who just happens to be in the wrong place at the wrong time (Frayer 1997). In contrast to this scenario, two mass graves were found in 1908 at Offnet in Germany, revealing the evidence of a massacre in which 37 individuals had been slaughtered; only the skulls and cervical vertebrae had been buried. This site represents the first clear evidence of a mass murder in prehistory, but why the 37 Mesolithic people, mainly women and children, were massacred and buried in such an unusual fashion remains unknown.

According to Constandse-Westermann *et al.* (1984) only very few cases indicate that the cause of death may have been related to direct interpersonal violence in the Western European Mesolithic. One of them was found in a *c.*7000-year-old grave in Vedbæk in Denmark in which a 1-year-old child lay between a male to the right and a female to the left. An arrow was embedded in one of the man's vertebrae. He had most likely been shot, but we do not know why the woman and child were buried with him (Bennike 1985). The Danish early Neolithic skeleton clearly illustrates that such an event by no means was restricted to the Mesolithic (Bennike 1999). A male skeleton was found with two arrows lodged in the nasal region and in the sternum leaving no doubt that the injuries were fatal.

The same seems to be the case with an early Neolithic female skeleton who was found with a string around her neck (Bennike and Ebbesen 1986).

The Greenland Eskimos and the Norse

The two populations from the third and final period, *c.*500–1000 years ago, are the Norse (Scandinavians) and the Eskimos or Inuits in Greenland. One may say that the comparison between these two groups lies at a racial or ethnic level, for there is no doubt that they are both genetically and culturally different. These two populations lived under almost identical geographical and climatic conditions for 500 years, in a country covered with ice and snow and with long, dark and cold winters. Genetically, they belonged to two of the main ethnic groups, the mongoloids and the caucasoids. Culturally, they belonged to different complexes. Whereas the Norse mainly immigrated to Greenland from Iceland and Norway around AD 950, the Eskimos are believed to have immigrated to Greenland in several waves, the first around 2000 BC. While the Eskimos as an ethnic group continued to survive in the very harsh climate, the Norse disappeared, for unknown reasons, from the area around AD 1450, some 500 years after their arrival. In fact, the fate of the population and their survival pattern outside Greenland is unknown, but later Eskimo skeletons do *not* seem to reflect that the two populations mixed (Jørgensen 1953). When Greenland was rediscovered by the Danes in the eighteenth century, no caucasoid morphology was noted among the Eskimo population, nor do the Norse skeletons found in Greenland exhibit any Eskimo morphology. Eskimos and Europeans have distinctly different cranial morphologies. The shape of Eskimo teeth, such as the marked shovel shape of the incisors, the long enamel extensors and the absence of large carabellis cusps, are part of the typical genetic pattern of the Eskimo dentition, whereas the opposite is typical for the caucasoid pattern of North Europe. Both from a biological and a cultural point of view, the Eskimos and Norse were very different, even though they belonged to the same species and subspecies of *Homo sapiens sapiens*.

The Eskimos and Norse are the only two populations who ever settled on the largest island in the world, Greenland. These two populations are extremely interesting with regard to survival, as the Eskimos survived whereas the Norse who settled around AD 950 disappeared for some unknown reason around 400–500 years later. Greenland is still inhabited by Eskimos, but their gene pool now encompasses some caucasoid genes. The European cultural pattern has also become part of the Greenlandic lifestyle now – for better or for worse.

The Norse

Skeletal remains of 457 Norse individuals have been found in Greenland. Several anthropological studies have been carried out in order to determine the cause of their disappearance during the fourteenth and fifteenth centuries. One study from the beginning of the twentieth century concluded that the Norse died out as a consequence of inbreeding (Hansen 1924). This assumption was based on the presence of degenerative arthritis in the joints and the spine. Furthermore, the deformations on a number of pelvic bones were interpreted as having been caused by rickets, vitamin-D deficiency. It was later demonstrated that degenerative arthritis is rather common in almost all skeletal samples with individuals above a certain age, and therefore cannot be attributed to inbreeding (Fischer-Møller 1942). The deformations of the pelvic bones were ultimatively proven to be post-mortem changes, probably due to the pressure of the soil. It also seems unusual to find vitamin-D deficiency in skeletons in areas where the middens that the Norse left behind provide ample evidence of oily fish and seals.

It has also been suggested that the Norse were attacked by the Eskimos and driven from Greenland, but there is no proof. One grave in a Norse churchyard at Tjodhilde provides evidence of a battle, but it may just as well have been the Norse who fought among themselves. The grave contains the skeletal remains of 13 inviduals, buried after they had almost decomposed. Several of the skulls have lesions produced by sharp weapons.

In the beginning of the fourteenth century the temperature dropped in Greenland. As the Norse more or less maintained their traditional subsistence economy which was mainly dependent on agriculture and sheep-raising, they would have been more vulnerable to the change in climate than the Eskimos were. One theory suggests that the Norse died of starvation, but the skeletal remains and the many archaeological finds do not bear evidence of such a dramatic event.

A study by Lynnerup (1998) reveals that Norse skeletons do not exhibit more pathological changes than their contemporary medieval counterparts in Scandinavia. Whether the Norse were eradicated by the plague or other sudden outbreaks of infectious diseases cannot be established, as neither leaves any revealing marks on the bones.

The Norse stature seems to have decreased over their 400–500 years in Greenland, although not significantly. The teeth are more worn and the ^{13}C content seems to indicate that the marine part of the diet increased (Arneborg *et al.* 1999). Furthermore, there seems to be an increase in the number of middle ear infections (Lynnerup 1998). As suggested by Lynnerup, the departure of the Norse may be less dramatic, gradually leaving Greenland as life conditions became worse. If this particular scenario reflects the past, the settlement's demise appears less spectacular than in other interpretations. The Norse could have moved to Greenland because of an anticipated gain and the opportunity to acquire land. Furthermore, some degree of population pressure following their rapid exploitation of Iceland may also have encouraged them to move on. They moved away when such opportunities presented themselves elsewhere. It would be surprising, in the light of almost general, widespread demographic changes and overall depopulation in Norway, Iceland, England etc., if a remote and already economically vulnerable settlement as the Norse settlement in Greenland should not decline (Lynnerup 1998). The survival of the Eskimos in Greenland is clear, and their strategy must have included the maintenance of their cultural pattern which was adapted to arctic conditions down to the smallest detail. This continued until the eighteenth century, when the caucasoids, as elsewhere, disturbed the ecological balance the Eskimos were so dependent on and had learnt to respect through many generations.

As for the disappearance of the Greenland Norse, it seems that their survival strategy, in the circumstances an intelligent one, was to move on from Greenland as the conditions of life deteriorated beyond endurance.

Acknowledgements

This chapter is based on a lecture given at the eleventh European Anthropological Association Meeting in Jena 1998 and in a revised form at a symposium of the Society for the Study of Human Biology in Durham, England 1999. Susan Peters is cordially thanked for her English revision of the manuscript and Verner Alexandersen for his critical reading of the manuscript.

References

Ammermann, A. J. and Cavalli-Sforza, L. L. (1984) *The Neolithic Transition and the Genetics of Populations in Europe*, Princeton, NJ: Princeton University Press.

Arneborg, J., Heinemeier, J., Lynnerup, N., Nielsen, H. L., Rud, N. and Sveinbjörndóttir, A. E. (1999) 'Change of diet of the Greenland Vikings determined from stable carbon isotope analysis and C-14 dating of their bones', *Radiocarbon*, 41, 157–68.

Barbujani, G., Pilastro, A., Domenico, S. de and Renfrew, C. (1995) 'Genetic variation in the North Africa and Eurasia: Neolithic demic diffusion vs. Paleolithic colonization', *American Journal of Physical Anthropology*, 95, 137–54.

Bennike, P. (1985) 'Palaeopathology of Danish skeletons: a comparative study of diseases, injuries and demography', Copenhagen: Akademisk Forlag.

Bennike, P. (1999) 'The Early Neolithic Danish bog finds: a strange group of people!', in Coles, B., Coles, J. and Jørgensen, M. S. (eds) *Bog Bodies, Sacred Sites and Wetland Archaeology*, Exeter: WARP Occasional Paper, Short Run Press, pp. 27–32.

Bennike, P. and Ebbesen, K. (1986) 'The bog find from Sigersdal: human sacrifice in the Early Neolithic', *Journal of Danish Archaeology*, 5, 85–115.

Berger, T. D. and Trinkaus, E. (1993) 'Late Archaic traumatic injuries activity and/or a bias in the record', *American Journal of Physical Anthropology*, 16, 56.

Churchill, S. E. (1998) 'Cold adaption, heterochrony, and Neanderthals', *Evolutionary Anthropology*, 7, 46–61.

Constandse-Westermann, T., Newell, R. R. and Meiklejohn, C. (1984) 'Human biological background of population dynamics in the Western European Mesolithic: human palaeontology', *Proceedings of the Koninklijke Nederlandse Akademie van Wetenschappen*, Series B, 87, 139–223.

Crubézy, E. and Trinkaus, E. (1992) 'Shanidar 1: a case of Hyperostotic Disease (DISH) in the Middle Paleolithic', *American Journal of Physical Anthropology*, 89, 411–20.

Dastugue, J. (1982) 'Les maladies des nos ancetres', *La Researche*, 13, 980–8.

Dastugue, J. and de Lumley, M-A. (1976) 'Les maladies des hommes prehistorique du Paleolithique et du Mesolithique', in de Lumley, H. (ed.) *La Prehistoire française 1*, Paris: Centre National de la Researche Scientifique, pp. 612–22.

Demes, B. (1987) 'Another look at an old face: biomechanics of the Neanderthal facial skeleton reconsidered', *Journal of Human Evolution*, 16, 297–305.

Fischer-Møller, K. (1942) 'The Medieval Norse settlements in Greenland: anthropological investigations', *Meddelelser om Grønland*, 89, 1–82.

Formicola, V., Pontrandolfi, A. and Svoboda, J. (1998) 'The Upper Palaeolithic triple burial of Dolni Vestonice: pathology and funeral behavior', *American Journal of Physical Anthropology, Supplement*, 26, 83.

Frayer, D. W. (1981) 'Body size, weapon use, and natural selection in the European Upper Paleolithic and Mesolithic', *American Anthropologist*, 83, 57–73.

Frayer, D. W. (1984) 'Biological and cultural change in the European Late Pleistocene and Early Holocene', in Smith, F. H. and Spencer, F. (eds) *The Origins of Modern Humans: a world survey of the fossil evidence*, New York: Alan R. Liss, pp. 211–50.

Frayer, D. W. (1997) 'Ofnet: evidence for a Mesolithic massacre', in Martin, D. L. and Frayer, D. W. (eds) *Troubled Times: violence and warfare in the past. War and Society, 3*, Gordon & Breach, pp. 181–216.

Frayer, D. W., Macchiarelli, R. and Mussi, M. (1988) 'A case of Chondodystrophic Dwarfism in the Italian Late Upper Paleolithic', *American Journal of Physical Anthropology*, 75, 549–65.

Hansen, F. C. C. (1924) 'Anthropologia Medico-Historica Groemlamdiae Antoquae. I Herjolfsnes', *Meddelelser om Grønland*, 67, 293–547.

Harding, R. M., Rösing, F. W. and Sokal, R. R. (1990) 'Cranial measurements do not support Neolithization of Europe by demic expansion', *Homo*, 40, 45–58.

Hoffecker, J. F. (1999) 'Neanderthals and modern humans in Eastern Europe', *Evolutionary Anthropology*, 7, 129–41.

Jackes, M., Lubell, D. and Meiklejohn, C. (1997) 'Healthy but mortal: human biology and the first farmers of western Europe', *Antiquity*, 71, 639–58.

Jørgensen, J. Balslev (1953) 'The Eskimo skeleton', *Meddelelser om Grønland*, 146, 2, Copenhagen: C. A. Reitzels Forlag.

Klein, R. G. (1989) *The Human Career: human biological and cultural origins*, Chicago: University of Chicago Press.

Klein, R. G. (1995) 'Neanderthals and modern humans in West Asia: a conference summary', *Evolutionary Anthropology*, 4, 187–93.

Klima, B. (1987) 'A triple burial from the Upper Palaeolithic of Dolni Vestonice', *Journal of Human Evolution*, 16, 831–5.

Krings, M., Stone, A., Schrintz, R. W., Kranitzki, H., Stoneking, H. and Pääbo, S. (1997) 'Neanderthal DNA sequences and the origin of modern humans', *Cell*, 90, 19–30.

Lahr, M. M. (1996) *The Evolution of Modern Human Diversity*, Cambridge: Cambridge University Press.

Larsen, C. S. (1995) 'Biological changes in human populations with agriculture', *Annual Review of Anthropology*, 24, 185–213.

Lynnerup, N. (1998) *The Greenland Norse: a biological-anthropological study*, Meddelelser om Grønland, Man and Society 24, Copenhagen: The Danish Polar Centre.

Meiklejohn, C. and Zwelebil, M. (1991) 'Health status of the European populations at the agricultural transition and the implications for the adoption of farming', in Bush, H. and Zwelebil, M. (eds) *Health in Past Societies*, BAR, Int. Series 567, pp. 129–45.

Mithen, S. (1996) *The Prehistory of the Mind*, London: Thames & Hudson.

Rak, Y. (1987) 'The Neanderthal; a new look at an old face', *Journal of Human Evolution*, 15, 151–64.

Richards, M., Côrte-Real, H., Forster, P., Macaulay, V., Wilkinson-Herbots, H., Demaine, A., Papiha, S., Hedges, R., Bandelt, H-J. and Sykes, B. (1996) 'Paleolithic and Neolithic lineages in the European mitochondrial gene pool', *American Journal of Human Genetics*, 59, 185–203.

Roper, M. K. (1969) 'A survey of the evidence for intrahuman killing in the Pleistocene', *Current Anthropology*, 10, 427–59.

Ruff, C. B. (1992) 'Biomechanical analyses of archaeological human skeletal samples', in Saunders, S. and Katzenberg, K. (eds) *Skeletal Biology of Past Peoples: research methods*, New York: Wiley-Liss, pp. 37–58.

Ruff, C. B., Larsen, C. S. and Hayes, W. C. (1984) 'Structural changes in the femur with the transition to the agriculture on the Georgia coast', *American Journal of Physical Anthropology*, 64, 125–36.

Straus, L. G. (1995) 'The Upper Paleolithic in Europe: an overview', *Evolutionary Anthropology*, 4, 4–16.

Trinkaus, E. (1985) 'Pathology and the posture of the Chapelle-aux Saints Neandertal', *American Journal of Physical Anthropology*, 67, 19–41.

Trinkaus, E. (1995) 'Neanderthal mortality patterns', *Journal of Archaeological Science*, 22, 121–42.

Trinkaus, E. and Zimmermann, M. (1982) 'Trauma among the Shanidar Neanderthals', *American Journal of Physical Anthropology*, 57, 61–76.

Vallois, H. V. (1961) 'The social life of early man: the evidence of skeletons', in Washburn, S. L. (ed.) *The Social Life of Early Man*, Chicago: Aldine, pp. 214–35.

Wolpoff, M. (1998) 'Concocting a divisive theory', *Evolutionary Anthropology*, 7, 1–3.

Wolpoff, M. and Caspari, R. (1997) *Race and Human Evolution*, New York: Simon & Schuster.

9 'Observe: our noses were made to carry spectacles, so we have spectacles'

(Dr Pangloss, from *Candide* by Voltaire)

Christopher J. Knüsel

The Panglossian Paradigm

Dr Pangloss espoused a fatalistic philosophy that justified inequalities by arguing that they were preordained and the best that were possible. Therefore, all features have a function and that function is to serve the best and present purpose. Hence, the Panglossian Paradigm states 'all is for the best' (Voltaire, *Candide*, p. 20) in this, the best of all possible worlds. When applied to morphology, Gould and Lewontin (1979) have identified this reasoning with the 'sign' theory of morphology. The causes of morphological features are to be found in their function, often without reference to their histories. Proponents of the sign theory see structures only as a means of tracing lineages, but they do not consider these structures as useful designs for existence. In other words, they privilege genetic causes to explain the appearance of morphological features. They associate morphological consequence with an often loosely defined genetic 'predisposition' seen to be of benefit to the individual organism, allowing it to adapt to changing circumstances.

Gould (1971) reminds us that no explanation of form is complete without reference to the physical forces that contribute to the appearance of anatomical elements. He identifies three types of adaptation, only one of which can be called Darwinian adaptation, that heritable form of adaptation that involves selection upon genetic variation. The other two, often confused with Darwinian adaptation, are physiological adaptation, encompassing plastic changes which accommodate the organism to a particular environmental extreme located within its range; and cultural adaptation, which is heritable in that it is learned and passed from one generation to another. Bock and von Wahlert (1965: 284) discuss physiological adaptation in the following manner: 'We regard physiological adaptation as a special case of the general principle that the phenotype is an expression of the genotype in a particular environment'. This chapter will demonstrate that Gould and Lewontin's (1979) paper is only one of the more recent – albeit one of the most eloquent – calls for a consideration of causes other than genetic ones to explain morphological change.

Plasticity: definitions and historical context

Plasticity, 'the capability of being moulded' (Roberts 1995: 1), relates to the malleability of organ systems in the face of variable climates, topography, diet, and physical activity (function). In modern use, plasticity refers to specific adaptations made during growth and development that are reversible and not inherited (Schell 1995: 219–20); they are thus part of the phenotype. Plastic changes are physiologically mediated responses that may not be adaptive for the organism, so plasticity and adaptation are not synonymous. Plasticity, though, is of major importance to the study of both accommodation and adaptation in human populations (see Frisancho 1993; Ulijaszek 1997; Roberts 1995 for reviews of these concepts). In an earlier review article on the subject of human adaptability, Lasker (1969) defined adaptation as the 'change by which organisms surmount the challenges of life'. Accommodation is understood to mean physiological response to environmental stimuli that are not entirely successful (Frisancho 1993: 7). This definition of accommodation seems to encompass Schell's (1995: 220) rewriting of Lasker's original definition; namely that adaptation consists in the 'change[s] an organism makes to surmount the challenges of life'. In short, individuals exhibit a range of responses to environmental stimuli, some of which may be better than others in certain circumstances. Accommodation permits organisms to survive, while adaptation allows them to thrive. Thus plastic change may not, necessarily, be for the best.

Studies of plastic alterations of the body were rare until relatively recently, in part because biometrical studies of the human body developed in an atmosphere of racial typology in the nineteenth and twentieth centuries. Essentially, early researchers emphasised what they interpreted to be innate differences among human groups based on physical appearance (see Gould 1981; Marks 1995). More recently, biometry has been frequently used to characterise biological variation and function among and within human groups, past and present. One of the earliest applications of the latter was by Franz Boas (1912) in his study entitled *Changes in Bodily Form of Descendants of Immigrants*, which was presented to the United States Congress on 16 December 1909. This research demonstrated that American-born descendants of immigrants differed from those who were born overseas. Boas attributed these differences to the altered environment in which emigrants grew up. In other words, he attributed these changes in bodily and facial form to nurture rather nature, as there had been no genetic admixture in his samples. In passing he warned:

> As long, then, as we do not know the causes of the observed changes, we must speak of plasticity (as opposed to permanence) of types, including in the terms changes brought about by any cause whatever – by selection, by changes of prenatal or postnatal growth, or by changes in the hereditary constitution of the individual. It is quite arbitrary to restrict plasticity to the last-named cause.
>
> (Boas 1912: 53)

The origins of modern interest in plastic change

A desire to better understand osteoporosis stimulated clinical interest in alterations in bone density and bone mineral content (Montoye *et al.* 1976; Huddleston *et al.* 1980; Talmage *et al.* 1986), as a consequence of immobilisation (Uhthoff and Jaworski 1978; Prince *et al.* 1988) and weightlessness (Whedon 1984), as well as in response to prolonged strenuous activity (see p. 190). In an early clinical study, Nilsson and Westlin (1971) found that Olympic athletes and physically active individuals possessed greater bone density than those who were not, and that the stature or weight of the individuals did not influence these differences. Individuals who exercised moderately showed greater bone density than those who did not (Nilsson and Westlin 1971), although consistent runners showed statistically significant bone density increase while those who ran inconsistently experienced gain but not to significant levels (Williams *et al.* 1984). Huddleston *et al.* (1980) compared the bone mineral mass of the radii of 35 professional lifetime tennis players, aged 70 to 84 years, with those of non-athletes and found that the dominant (i.e. playing) forearm (1.37 g/cm) of the tennis players had greater bone mineral content than did their non-dominant forearms (1.23 g/cm) and of the dominant forearms of a non-athletic group. The results of these studies suggested that bone mass and density could be increased through exercise and that, if the metabolic mechanisms could be harnessed, fragile bones might be avoided.

Previously, in the late 1940s, 1950s and 1960s physicians identified asymmetrical limb bone development during their diagnosis and treatment of injuries suffered by professional athletes (Bennett 1941, 1947, 1959; Bateman 1962; Brewer 1962). Many subsequent studies of physically active, modern individuals record localised bone hypertrophy in response to strenuous activity, including the ulna in rodeo riders (Claussen 1982), the dominant (e.g. playing side) elbow joint in baseball pitchers (King *et al.* 1969; Watson 1973) and tennis players (Lewis 1971).

Among the highly physically active groups studied to date, professional tennis players, who normally train from a young age, have recorded some of the most exuberant bone responses in their dominant (e.g. playing side arms) when compared with their contralateral arm. Professional male tennis players have as much as 34.9 per cent more bone in their dominant arms; female professionals have as much as 28.4 per cent (Jones *et al.* 1977). The study sample consisted of individuals aged between 18 and 50 years (men) and 14 and 34 years (women), who had been playing tennis for an average of 18 and 14 years, respectively, from ages as young as 5 to 19 years of age (Ruff *et al.* 1994). Those who had played from a young age showed greater bone change (greater cortical bone area, total cross-sectional area, and torsional strength) than those who began playing in mid-adolescence (Ruff 1992).

Bone change is mediated through the activity of tendons and muscles of the body. Bone is considered to be anisotropic because it responds differently when loaded from different directions (e.g. it has a grain) (Currey 1984; Lanyon *et al.*

1982). Bone is thus best understood as a living and dynamic biological structure that forms and reforms in response to activity as the body grows and develops. The processes of bone adaptation are predictable according to biological laws, most fundamental of which is that of the German anatomist, Julius Wolff (1836–1902). Wolff's law states that 'the form of bone being given, the bone elements place or displace themselves in the direction of the functional pressure and increase or decrease their mass to reflect the amount of functional pressure' (Murray 1936: 134). In other words, bone forms in a manner such that its surface morphology, cross-sectional area, density and shape are best able to withstand the applied stress of repeated movement. Essentially, this relationship maintains the body's integrity throughout the lifetime of the individual (Rubin et al. 1990). Previous research into the differing responses of adult bone and that of juveniles has demonstrated that the greatest impact on bone morphology is made before physiological maturity (Kiiskinen 1977; Woo et al. 1981; Anderson 1996). Therefore, tasks that are commenced early in the lifetime of the individual will present the greatest osseous responses.

Posture greatly affects the loading of bone. In a review of cross-cultural postures, Hewes (1955) demonstrated that many are culturally determined and that, though the body is capable of a whole range of postures, some are preferentially used and others not. Those postures assumed are based not only on the type of task performed but also on the age and sex of individuals which, in turn, determines the time of their induction into adult status and activities (cf. Schlegel and Barry 1979–80; Murdock and Provost 1973). Therefore, postures, like activities, relate to specific cultural, technological and social circumstances, as well as biological ones.

Many times body weight is applied to skeletal elements through postural changes. Walking, for example, exerts two to three times body weight on the knee joint during heel strike (Nordin and Frankel 1989a: 127) and, on average, from four (women) to seven times (men) body weight at the hip just prior to toe-off (Nordin and Frankel 1989b: Figure 7–15). The ground reaction force (that force transmitted to the ground) at the mid-stance phase (where one foot is on the ground) of running is equivalent to 250 to 300 per cent of body weight; thereby, a runner absorbs the equivalent of 110 tons on each foot per mile covered (Renstrsm and Johnson 1985). Throwing produces joint reaction forces (the internal force acting on articular surfaces when a joint is loaded) approaching 90 per cent of body weight within the shoulder (Zuckerman and Matsen 1989), which can produce a speed of 45 m/sec (160 km/h) in a ball thrown by professional baseball pitchers (Fleisig et al. 1996). Activities involving throwing result in asymmetrical muscular hypertrophy (Figures 9.1 to 9.3). Javelin-throwing places between 2.8 and 6.6 times body weight on the leading foot (i.e. the contralateral one, opposite the throwing arm) when it strikes the ground, producing throws of between 69.40 and 88.14 m in elite athletes and about 30 m in novices (Morriss and Bartlett 1996). There are few studies documenting similar parameters for other, common activities and postures.

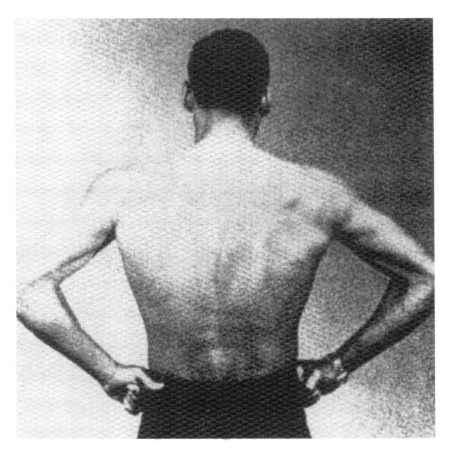

Figure 9.1 The hypertrophied musculature of a left-handed pitcher. Note especially the hypertrophy of the rotator cuff musculature covering the squamous portion of the scapula and the greater size of the arm, forearm and neck musculature of the dominant, left side of this individual (e.g. the throwing side). (Reprinted with permission from 'Analysis of the pitching arm of the professional baseball pitcher' by King, J. W., Brelsford, H. J. and Tullos, H. S. from *Clinical Orthopaedics and Related Research*, 1969, 67, pp. 116–23, Figure 1, p. 116.)

Controversy and context

Armed with the concept of plasticity and empirical evidence of the effects of activity on bone morphology, researchers have attempted to isolate particular morphologies indicative of specific activities. Several researchers have questioned the validity of these attempts (see a review of these in Knüsel 2000). The basic and undeniable problem is that different conditions can produce similar bone responses, be they the consequence of pathology, congenital anomaly, or acquired in the course of growth and development. These uncertainties have produced variable responses among researchers. The most damning come from those engaged

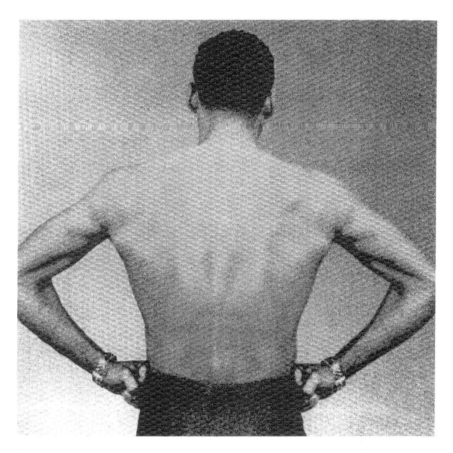

Figure 9.2 A mirrored image of the same left-handed pitcher made from his non-dominant, right side. (Reprinted and modified with permission from 'Analysis of the pitching arm of the professional baseball pitcher' by King, J. W., Brelsford, H. J. and Tullos, H. S. from *Clinical Orthopaedics and Related Research*, 1969, 67, pp. 116–23, Figure 1, p. 116.)

in studies of osteoarthritis (cf. Waldron 1994), a complex multi-factorial disorder that seems to have more to do with previous soft tissue injury, altered osseous architecture and consequent loss of fine neuromuscular control than with activity per se (Anderson 1984). More recently, Jurmain (1999) views some biocultural interpretations of activity as 'just-so stories' derived from ethnohistorical and historical information applied to explain the morphological patterns observed – in other words, susceptible to accusations of Panglossian reasoning. In order to understand the potential and pitfalls of identifying activity from morphology, a review of the historical development of activity-related studies shows that the shortcomings are as much a product of their origin as of their novelty, but that there is much to recommend their continued study.

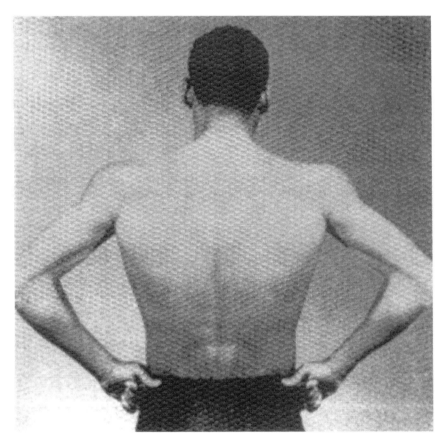

Figure 9.3 A mirrored image of the same left-handed pitcher made from his dominant, left side. A comparison with Figure 9.2 reveals the extent of the bilateral asymmetry of this professional athlete. (Reprinted and modified with permission from 'Analysis of the pitching arm of the professional baseball pitcher' by King, J. W., Brelsford, H. J. and Tullos, H. S. from *Clinical Orthopaedics and Related Research*, 1969, 67, pp. 116–23, Figure 1, p. 116.)

Early developments

Bernardo Ramazzini, Professor of Medicine at the Universities of Modena and Padua, published the first treatise on occupational disease, *De Morbis Artificium* (*Diseases of Workers*), in AD 1700 (Ramazzini 1964). Written in the form of a recommendation, it advised medical professionals to question their patients about the types of activities in which they engaged in order to assess physical ailments. He wrote: 'In medical practice, however, I find that attention is hardly ever paid to this matter, or if the doctor in attendance knows it without asking, he gives little heed to it, though for effective treatment evidence of this sort has the utmost weight' (Ramazzini 1964: 7). Based on their origin, Ramazzini (1964: xxvi) provided the following classification of these ailments: 'First, by the injurious character of the materials handled, which emit noxious vapours . . . secondly, by certain violent and irregular motions and unnatural postures of the body, by which the natural structure of the vital machine is impaired'. He differentiated between two categories of conditions, those relating to the materials with which people worked and those known as cumulative trauma disorders today. The latter are defined as 'a category of physical signs and symptoms due to chronic musculoskeletal injuries where the antecedents (causes) appear to be related to some aspect of repetitive work' (Putz-Anderson 1988: 15).

In many cases, though, Ramazzini noted only that those employed in a particular way developed similar conditions without understanding the specific cause, a problem that continues to the present day in occupational medicine. For example, he noted an association between breast cancer and nuns and thought it connected in some way to their celibate lives. Other, more easily perceptible ailments afflicted midwives (inflamed and ulcerated hands), stone cutters (asthmatic coughs), carpenters and other craftsmen who must stand in the course of their work (varicose veins), scribes and notaries (hand and arm fatigue), among others. Although many of these descriptions resonate with modern analogues, the observations upon which they are based are very much a product of Ramazzini's time, place and society. This is especially true when he describes workers in relation to their tools or places of work:

> Bakers suffer from yet another ailment; they all become bow-legged as anyone can see, i.e. their calves are bent outwards so that they resemble crabs or lizards. This is because in the country on both sides of the Po they work in the following way: Above a thick board or three-legged table they fasten a smooth block of wood of conical shape at the top of a three-legged frame in such a way that it can be rotated, then they place a great mass of dough underneath, drop down the wooden block from above and press it with all the strength of their arms and with their knees as well, while someone keeps turning over the dough. Since the articulation of the knees is not very strong, their legs are bowed outwards. For this trouble there is no remedy, for even when they are young and strong, they soon become bow-legged and, in course of time, lame.
>
> (Ramazzini 1964: 230)

These anecdotal references, although fascinating for the light they throw on the development of occupational medicine, are of somewhat limited use with regard to finding similar manifestations in undocumented skeletal remains. It is not clear what, if any, skeletal change is involved or how manifestations relating to a particular trade would be differentiated from similar ones relating to other conditions. Of more potential use are Ramazzini's more ergonomic descriptions because they have more immediately perceived modern parallels. After noting that women remain erect when they carry loads on their heads in order to avoid collapsing under the weight of the burden, he posits that men act otherwise when carrying their more bulky burdens on their backs rather than solely on their shoulders because:

> the load would press on the collarbone, which is a small bone, and the weight would fall on the middle more than on the extremities, so that it might easily be broken; whereas when he is stooping and leaning forward, the weight rests on the shoulder-blade which is a large, broad, strong bone, so that the pressure is less painful. . . . Now, when a load is adjusted on the shoulder of a porter as he leans forward, the part on which it lies is stronger, and moreover the pressure is distributed over more parts of the body than if he held himself erect; and this applies to any load, e.g. wood . . . a sack of wheat; hence it is easier to support the weight; and therefore porters . . . lean forward and stick their buttocks out behind so as to keep the center of gravity in a straight line.
>
> (Ramazzini 1964: 315)

Ramazzini's treatise was transmitted to Britain through a translation published in 1705, entitled *A Treatise on the Disease of Tradesmen*, which Charles Turner Thackrah encountered during his medical training at Guy's Hospital, London, and inspired his 1831 treatise on *The Effects of the Principal Arts, Trades, and Professions, and of Civic States and Habits of Living, on Health and Longevity* (Kennedy 1989). Fifty years later, Arbuthnot Lane, Senior Demonstrator in Anatomy at Guy's Hospital, was among those stimulated by a similar interest in occupations and their effect on the body. During the 1880s, Lane published a series of articles documenting several instances of anatomical alterations he considered to be associated with the pursuit of strenuous activities and, in some cases, with particular professions. He published on those individuals for whom he could obtain some form of documentation relating to the labour-history of the individual. In the 'Anatomy and physiology of the shoemaker', Lane (1888a: 593) writes, 'Having concluded, from a careful examination of the changes which the body presented, that the man had been a shoemaker, I wrote to the medical officer of the infirmary in which he died for any information he could give me, and he kindly informed me that the man was entered on the books as a shoemaker' In other instances, though, Lane was less successful in obtaining documentary support; he had to *surmise* what sort of pursuit was linked to the anomalies he noted. In his 'A remarkable example of the manner in which pressure-changes in the skeleton may reveal the labour-history of the individual', Lane (1887: 397) comments on the difficulties encountered in trying to get labour histories of the individuals he dissected. In the case of his ship's trimmer his documentation was obtained quite circuitously from comments made by various men employed in shipping to conclude that shipkeepers, the last recorded profession of one individual, were often carpenters or trimmers (essentially coal shovellers) before becoming shipkeepers. The circumstances of Lane's efforts are little different from those facing researchers today, although global economic and social transformations are quickly reducing the number of traditional trades and their practitioners, leaving us with few living examples of or those knowledgeable about earlier professions.

From the comments in these publications, it is clear that Lane realised that there was a certain redundancy in the expression of anatomical modification in relation to the movements associated with labour. Lane (1888a: 626) writes: 'we cannot observe and define too accurately the various modifications in form which the normal bones undergo under the influence of the groups of movements which compose some of the more routine forms of labour, because they represent the exaggeration . . . of the results [of] one single movement . . . habitually performed in [a] particular labour'. He also realised that the duration and intensity of movement were implicated in the expression of these changes (Lane 1888a: 626–7) and that activities commenced early in life were most likely to produce anatomical modification, specifically, in relation to his ship's trimmer (Lane 1887). Lane (1887) also noted that some elements were hypertrophied (enlarged) as a result of muscular exertion and ascribed the observed changes associated with the 'so-called disease rheumatoid arthritis' to frequent exercise and the enormous pressures exerted on the body over a long period of time. Following a similar intuition about aetiology, a number of researchers have addressed the relationship between what, today, is called osteoarthritis and activity.

Lane's writings have continued to influence the work of medical practitioners to the present day. One product of this interest is Hunter's *Diseases of Occupations* (1969), which provided an update of Ramazzini's documentation of occupationally related ailments encountered in postindustrial society. For the skeletal biologist, it suffers from the same drawbacks as its predecessor in that there is little information provided on the skeletal manifestations of the disorders reviewed. Among those conditions that are accompanied by skeletal manifestations are acquired scoliosis due to burden carrying on one shoulder (with a figure of Lane's scoliotic spine: Figure 294), an assessment which harkens to Ramazzini's mention of one side of the body being stronger in such individuals (see p. 186). Hunter (1969: 802–4) noted similar changes in boot welders who perform an asymmetrical bending in the performance of their work. He also recorded dental wear associated with upholsterers who hold nails in their teeth (Figure 293). As with Ramazzini's treatise, these are anecdotal occurrences that are very likely found in others performing similar tasks in a variety of professions and vocations.

About the same time and, again, drawing inspiration from Lane's work, Wells (1967) described the occurrence of three individuals exhibiting spinal deformities in a small group of eight individuals preserving vertebral columns from the cemetery associated with St Michael-at-Thorn, Ber Street, Norwich, a medieval foundation. One of these, burial 10, the remains of a middle-aged man, exhibited a kyphosis and lower thoracic scoliosis accompanied by vertebral osteophytes, posterior ankylosis of the third and fourth cervical vertebrae, bilateral osteophytosis of the ischial tuberosities, and medial curvature of the distal portion of the fibulae. Because Norwich had been a thriving textile centre in the medieval period and St Michael's was located in a parish of artisans in earlier times, Wells considered that the osteophytosis of the ischial tuberosities was due to an inflammatory bursitis commensurate with 'weaver's bottom' and that the bowing of the fibulae resulted from sitting cross-legged in an attitude often associated with tailors (thus employing the eponymous *M. sartorius*). Images of medieval weavers, who employed low stools and horizontal looms, adopted a similar posture (see Basing 1990: compare Plate VII, p. 87). Although an elegantly constructed argument, Wells laments the lack of analogous studies of tailors' legs in order to provide further substantiation of his intuition. If Wells is correct, then this man's remains reflect the shift from weaving being a trade associated with female labour and the warp-weighted loom in the pre-Conquest period (prior to the eleventh century) to one dominated by the larger, more powerful horizontal loom and male labour in the later medieval and early-modern periods (Walton 1991: 346).

In the years since these treatments, and as reflected by Kennedy's (1989) review article, case studies of individual activity-related alterations have continued to appear. In many instances, however, researchers have been tempted to use such features in a typological sense, rather than as a means to understand the process by which these modifications occur and under what circumstances. As noted by Angel (1964) with regard to a group of non-metric traits of the proximal femur and, later, by Trinkaus (1975) with regard to the features considered to result from squatting, skeletal traits of the pelvic girdle and lower limbs appear to reflect rather generalised patterns of movement and posture that cannot be associated, solely, with a single activity.

As with Ramazzini's original cases, such treatments, even typologically based ones, have also often served more to emphasise that such studies can contribute to understandings of the human form in an intellectual climate that has been dominated by what Bock and von Wahlert (1965: 270) have termed 'the study of pure form divorced from function'. They write:

> our understanding of biological adaptation contains certain serious limitations which may be traced back to the prevailing philosophy accepted by anatomists in the last century. The most pertinent element of this philosophy is the postulate that morphology should be a study of pure form divorced from function. Morphological features were treated as geometrical units that changed during ontogeny and phylogeny according to rigid and often rather 'biologically abstract' mathematical laws. Structures were not regarded, as they should be, as biological features functioning together as integral parts of the whole organism. Nor were the changes in these structures during ontogeny and phylogeny regarded as modifications in response to alterations in relationship between the form-function complex and the environment.
>
> (Bock and von Wahlert 1965: 269–70)

This prevailing view continues to impinge on the application of these understandings to interpretations of skeletal morphology. This neglect may relate to an hereditarian assumption that was also intimated by Lane, himself, at the end of his treatment of the shoemaker when he invokes an 'hereditary factor' in the production of anatomical changes related to occupation in offspring and, specifically, those of labourers (Lane 1888a: 627). Like some of his more sagacious successors, though, Lane expresses ambivalence as to the cause of the alterations he observed. In doing so, he adds another dimension to Ramazzini's second category concerning violent and irregular motions. In the introduction to his shoemaker, he wrote:

> I believe that it is by the careful observation of the changes in form and structure which bones, joints, and muscles undergo, when exposed to the influence of a series of definite movements, that we shall obtain an accurate insight into the various factors that determine not only the variations in the character of these structures in individuals of the same race, but also in members of the different races of man and of the quadrumana, together with the manner in which the factors evolved by an alteration in habits, resulting from a change in the surroundings, experience, &c., of the individual and so produced new types.
>
> (Lane 1888a: 593)

Here, although relying on a now largely defunct typological approach to human variation, Lane anticipates today's concept of physiological adaptation, non-pathological changes, which permit individuals to adjust to their environment through bodily modification. He states this more plainly in another article from the same year (Lane 1888b: 222), writing 'I am rapidly collecting material by means of which I think that I shall be able to prove . . . that the form of the head, like that of the components of the trunk, and probably also of the limbs, will undergo a change in response to a variation in its mechanical relations to its surroundings'. This perspective converges with an anthropological interest in activity variants of past human groups and presages a shift away from the postindustrial concern with the effect of profession on the human body.

Conclusions

Despite early work on the physical manifestations of movement and activity, one that predates the synthesis of Darwinian evolution with modern genetics in the 1930s, research into activity-related change remains a minor intellectual companion to studies involving the genetic basis of form and behaviour. This situation likely arose because of the anecdotal nature of many early studies. These were heavily influenced by the amount of contextual information about individuals that could be gathered, often from the comments of knowledgeable informants. Lacking population comparisons, these studies did not address the replicability of the relationship between the observations of skeletal alteration and activity. Many of the early observations, though, rather then being of little value due to this lack, have great empirical value as hypotheses to be tested on a population level. The obstacles to this type of research today are similar to those that influenced earlier ones. First, in an atmosphere – both scientific and popular – that continues to ascribe behaviour and physical differences to underlying genetic differences, such studies seem much less glamorous and headline-grabbing than those to do with genetic breakthroughs, real or otherwise. Due to the social conditioning of millennia of increasing social inequality within human groups and the ubiquity of official documents that divide peoples into arbitrary (and biologically nebulous) groups based on profession, class, visible physical differences and ethnic origin, people today are as likely as their forebears to accept that some are inherently different from others, that this has always been so, and will continue to be the case. Similar perceptions held on an official level stymie research into the acquired component of human potential and privilege those of supposed innate predispositions.

On a purely logistical level, studies of plastic change in response to activity are also hampered by the rapid loss of traditional crafts and craftsmen (cf. Dutour 1992), who could be the focus of future research. Studies of professional athletes or strenuously active military personnel (for example, see Stirland 1998) demonstrate the extremes of osseous response to activity. The recent advent of these activities and their social context make them of less value for specific comparisons with those performing crafts in the past, however. More specific analogues would come from those engaged in the fabrication of objects from stone, wood, horn, or metal, in the manufacture of ceramics and textiles, and in the use of domestic animals such as horses. Each of these materials has played a role in transformations of labour and social organisation in past societies (see, for example, Rowlands 1971; Sherratt 1983; Anthony 1986; Peregrine 1991; Sherratt and Sherratt 1993; Shennan 1999). Various population-based studies have demonstrated their use in comparing general levels of strenuous activity and mobility between groups in the context of major subsistence shifts (reviewed in Larsen 1997: 195–225; Jurmain 1999: 231–59; Knüsel 2000). Similar studies of time-successive early hominid groups have also elucidated the osseous impact of strenuous activity in the course of human evolution and its influence on bilateral asymmetry and hand preference (see Ruff *et al.* 1993, 1994; Trinkaus *et al.* 1994). Far fewer studies, though, have addressed similar developments in the context of major social transformations in later prehistory, protohistory and history (see Larsen and Ruff 1994, for one example). Studies of these social circumstances would be more in the vein of those of Ramazzini and Lane that examined the relationship between labour and its effect on the body within its social context. Such studies would lend themselves well to the critique of protohistoric and historical documents that purport to relate the circumstances of these social transformations. These studies have the potential to help find our 'apparently lost intellectual compass' (see comments in Jurmain 1999: 266–7) and to close the rifts between population-based (and processual?) skeletal biologists and individualistically oriented post-processualist archaeologists and anthropologists. To rephrase Dr Pangloss: 'Observe: our noses carry spectacles, so we have different noses'. Moreover, those that wear spectacles are able to see their world more clearly and this, in turn, alters the way they interact with it.

References

Anderson, J. A. D. (1984) 'Arthrosis and its relation to work', *Scandinavian Journal of Work and Environmental Health*, 10, 429–33.

Anderson, J. J. B. (1996) 'Development and maintenance of bone mass through the life cycle', in Anderson, J. J. B. and Garner, S. C. (eds) *Calcium and Phosphorus in Health and Disease*, Boca Raton, FL: CRC Press, pp. 265–88.

Angel, J. L. (1964) 'The reaction area of the femoral neck', *Clinical Orthopaedics*, 32, 20–142.

Anthony, D. W. (1986) 'The "Kurgan Culture" Indo-European origins, and the domestication of the horse: a reconsideration', *Current Anthropology*, 27, 291–313.

Basing, P. (1990) *Trades and Crafts in Medieval Manuscripts*, London: The British Library.

Bateman, J. E. (1962) 'Athletic injuries about the shoulder in throwing and body-contact sports', *Clinical Orthopedics*, 23, 75–83.

Bennett, G. E. (1941) 'Shoulder and elbow lesions of the professional baseball pitcher', *Journal of the American Medical Association*, 117, 510–14.

Bennett, G. E. (1947) 'Shoulder and elbow lesions distinctive of baseball players', *Annals of Surgery*, 26, 107–10.

Bennett, G. E. (1959) 'Elbow and shoulder lesions of baseball players', *American Journal of Surgery*, 98, 484–92.

Boas, F. (1912) 'Changes in bodily form of descendants of immigrants', *American Anthropologist*, 14, 530–63.

Bock, W. J. and von Wahlert, G. (1965) 'Adaptation and the form-function complex', *Evolution*, 19, 269–99.

Brewer, B. J. (1962) 'Athletic injuries: musculotendinous unit', *Clinical Orthopedics*, 23, 30–8.

Claussen, B. F. (1982) 'Chronic hypertrophy of the ulna in the professional rodeo cowboy', *Clinical Orthopaedics and Related Research*, 164, 45–7.

Currey, J. D. (1984) *The Mechanical Adaptations of Bone*, Princeton, NJ: Princeton University Press.

Dutour, O. (1992) 'Activités physiques et squelette humain: le difficile passage de l'actuel au fossile', *Bulletin et Mémoires de la Société d'Anthropologie de Paris*, n.s. 4, 233–41.

Fleisig, G. S., Barrentine, S. W., Escamilla, R. F. and Andrews, J. R. (1996) 'Biomechanics of overhand throwing with implications for injuries', *Sports Medicine*, 21, 421–37.

Frisancho, A. R. (1993) *Human Adaptation and Accommodation*, Ann Arbor, MI: University of Michigan Press.

Gould, S. J. (1971) 'D'Arcy Thompson and the science of form', *New Literary History*, 2, 229–58.

Gould, S. J. (1981) *The Mismeasure of Man*, New York: W. W. Norton.

Gould, S. J. and Lewontin, R. C. (1979) 'The spandrels of San Marco and the Panglossian paradigm: a critique of the adaptationist programme', *Proceedings of the Royal Society of London B*, 205, 581–98.

Hewes, G. W. (1955) 'World distribution of certain postural habits', *American Anthropologist*, 57, 231–44.

Huddleston, A. L., Rockwell, D., Kuland, D. N. and Harrison, R. B. (1980) 'Bone mass in lifetime tennis athletes', *Journal of the American Medical Association*, 244, 1107–9.

Hunter, D. (1969) *Diseases of Occupations*, London: English University Press.

Jones, H., Priest, J. D., Hayes, W. C., Tichenor, C. C. and Nagel, D. A. (1977) 'Humeral hypertrophy in response to exercise', *Journal of Bone and Joint Surgery*, 59-A(2), 204–8.

Jurmain, R. (1999) *Stories from the Skeleton: Behavioral Reconstruction in Human Osteology*, Amsterdam: Gordon & Breach.

Kennedy, K. A. R. (1989) 'Skeletal markers of occupational stress', in Iscan, M. Y. and Kennedy, K. A. R. (eds) *Reconstruction of Life from the Skeleton*, New York: Alan R. Liss, pp. 129–60.

Kiiskinen, A. (1977) 'Physical training and connective tissues in young mice-physical properties of Achilles tendons and long bones', *Growth*, 41, 123–37.

King, J. W., Brelsford, H. J. and Tullos, H. S. (1969) 'Analysis of the pitching arm of the professional baseball pitcher', *Clinical Orthopaedics and Related Research*, 67, 116–23.

Knüsel, C. J. (2000) 'Bone adaptation and its relationship to physical activity in the past', in Cox, M. and Mays, S. (eds) *Human Osteology in Archaeology and Forensic Science*, London: Greenwich Medical Media, pp. 381–401.

Lane, W. A. (1887) 'A remarkable example of the manner in which pressure-changes in the skeleton may reveal the labour-history of the individual', *Journal of Anatomy and Physiology*, 21, 385–406.

Lane, W. A. (1888a) 'The anatomy and physiology of the shoemaker', *Journal of Anatomy and Physiology*, 22, 593–628.

Lane, W. A. (1888b) 'Can the existence of a tendency to change in the form of the skeleton of the parent result in the actuality of that change in the offspring?', *Journal of Anatomy and Physiology*, 22, 215–24.

Lanyon, L. E., Goodship, A. E., Pye, C. J. and MacFie, J. H. (1982) 'Mechanically adaptive bone remodelling', *Journal of Biomechanics*, 15, 141–54.

Larsen, C. S. (1997) *Bioarchaeology: interpreting behavior from the human skeleton*, Cambridge: Cambridge University Press.

Larsen, C. S. and Ruff, C. B. (1994) 'The stresses of conquest in Spanish Florida: structural adaptation and change before and after contact', in Larsen, C. S. and Milner, G. R. (eds) *In the Wake of Contact: biological responses to conquest*, New York: Wiley-Liss, pp. 21–34.

Lasker, G. W. (1969) 'Human biological adaptability', *Science*, 166, 1480–6.

Lewis, C. W. D. (1971) 'Who's for tennis?', *New Zealand Journal of Medicine*, 4, 21–4.

Marks, J. (1995) *Human Biodiversity: genes, race, and history*, New York: Aldine de Gruyter.

Montoye, H. J., McCabe, J. F., Metzner, H. L. and Garn, S. M. (1976) 'Physical activity and bone density', *Human Biology*, 48, 599–610.

Morriss, C. and Bartlett, R. (1996) 'Biomechanical factors critical for performance in the men's javelin throw', *Sports Medicine*, 21, 438–66.

Murdock, G. P. and Provost, C. (1973) 'Factors in the division of labor by sex: a cross-cultural analysis', *Ethnology*, 12, 203–25.

Murray, P. D. F. (1936) *Bones*, Cambridge: Cambridge University Press.

Nilsson, B. E. and Westlin, N. E. (1971) 'Bone density in athletes', *Clinical Orthopaedics and Related Research*, 77, 179–82.

Nordin, M. and Frankel, V. H. (1989a) 'Biomechanics of the knee', in Nordin, M. and Frankel, V. H. (eds) *Basic Biomechanics of the Skeletal System*, 2nd edn, Philadelphia, PA: Lea & Febiger, pp. 115–34.

Nordin, M. and Frankel, V. H. (1989b) 'Biomechanics of the hip', in Nordin, M. and Frankel, V. H. (eds) *Basic Biomechanics of the Skeletal System*, 2nd edn, Philadelphia, PA: Lea & Febiger, pp. 135–51.

Peregrine, P. (1991) 'Some political aspects of craft specialization', *World Archaeology*, 23, 1–11.

Prince, R. L., Price, R. I. and Ho, S. (1988) 'Forearm bone loss in hemiplegia: a model for the study of immobilization osteoporosis', *Journal of Bone and Mineral Research*, 3, 305–10.

Putz-Anderson, V. (1988) *Cumulative Trauma Disorders: a manual for musculoskeletal diseases of the upper limbs*, London: Taylor & Francis.

Ramazzini, B. (1964) *Diseases of Workers*, translated from the Latin text, *De morbis artificum*, of 1713, by Wilmer Cave Wright, New York: Hafner.

Renstrŝm, P. and Johnson, R. J. (1985) 'Overuse injuries in sports', *Sports Medicine*, 2, 316–33.

Roberts, D. F. (1995) 'The pervasiveness of plasticity', in Mascie-Taylor, C. G. N. and Bogin, B. (eds) *Human Variability and Plasticity*, Cambridge: Cambridge University Press, pp. 1–17.

Rowlands, M. J. (1971) 'The archaeological interpretation of metalworking', *World Archaeology*, 3, 211–25.

Rubin, C. T. McLeod, K. J. and Bain, S. D. (1990) 'Functional strains and cortical bone adaptation: epigenetic assurance of skeletal integrity', *Journal of Biomechanics*, 23, 43–54.

Ruff, C. B. (1992) 'Age changes in endosteal and periosteal sensitivity to increased mechanical loading', *38th Annual Meeting of the Orthopaedic Research Society*, 17, 532.

Ruff, C. B., Trinkaus, E., Walker, A. and Larsen, C. S. (1993) 'Postcranial robusticity in *Homo*, I: Temporal trends and mechanical interpretation', *American Journal of Physical Anthropology*, 91, 21–53.

Ruff, C. B., Walker, A. and Trinkaus, E. (1994) 'Postcranial robusticity in *Homo*, III: ontogeny', *American Journal of Physical Anthropology*, 93, 35–54.

Schell, L. M. (1995) 'Human biological adaptability with special emphasis on plasticity: history, development and problems for future research', in Mascie-Taylor, C. G. N. and Bogin, B. (eds) *Human Variability and Plasticity*, Cambridge: Cambridge University Press, pp. 212–37.

Schlegel, A. and Barry III, H. (1979–80) 'Adolescent initiation ceremonies: a cross-cultural code', *Ethnology*, 18–19, 199–210.

Shennan, S. (1999) 'Cost, benefit and value in the organization of early European copper production', *Antiquity*, 73, 352–63.

Sherratt, A. (1983) 'The secondary exploitation of animals in the Old World', *World Archaeology*, 15, 90–104.

Sherratt, S. and Sherratt, A. (1993) 'The growth of the Mediterranean economy in the early first millennium BC', *World Archaeology*, 24, 361–75.

Stirland, A. J. (1998) 'Musculo-skeletal evidence for activity: problems of evaluation', *International Journal of Osteoarchaeology*, 8, 354–62.

Talmage, R. V., Stinnett, S. S., Landwehr, J. T., Vincent, L. M. and McCartney, W. H. (1986) 'Age-related loss of bone mineral density in non-athletic and athletic women', *Bone and Mineral*, 1, 115–25.

Trinkaus, E. (1975) 'Squatting among the Neandertals: a problem in the behavioral interpretation of skeletal morphology', *Journal of Archaeological Science*, 2, 327–51.

Trinkaus, E., Churchill, S. E. and Ruff, C. B. (1994) 'Postcranial robusticity in *Homo*, II: humeral bilateral asymmetry and bone plasticity', *American Journal of Physical Anthropology*, 93, 1–34.

Uhthoff, H. K. and Jaworski, Z. F. G. (1978) 'Bone loss in response to long-term immobilisation', *Journal of Bone and Joint Surgery*, 60-B (3), 420–9.

Ulijaszek, S. J. (1997) 'Human adaptation and adaptability', in Ulijaszek, S. J. and Huss-Ashmore, R. (eds) *Human Adaptability: past, present, and future*, Oxford: Oxford University Press, pp. 7–16.

Voltaire (François-Marie Arouet) ([1758] 1947) *Candide or Optimism*, translated by John Butt, Harmondsworth: Penguin.

Waldron, T. (1994) *Counting the Dead: the epidemiology of skeletal populations*, Chichester: John Wiley.

Walton, P. (1991) 'Textiles', in Blair, J. and Ramsay, N. (eds) *English Medieval Industries*, London: Hambledon Press, pp. 319–54.

Watson, R. C. (1973) 'Bone growth and physical activity in young males', in Mazess, R. B. (ed.) *International Conference on Bone and Mineral Measurement (Chicago, Oct. 1973)*, Bethesda, MD: National Institutes of Health, pp. 380–5.

Wells, C. (1967) 'Weaver, tailor, or shoemaker? An osteological detective story', *Medical Biology Illustrated*, 17, 39–47.

Whedon, G. D. (1984) 'Disuse osteoporosis: physiological aspects', *Calcified Tissue International*, 36, S146–S150.

Williams, J. A., Wagner, J., Wasnich, R. and Heilbrun, L. (1984) 'The effect of long-distance running upon appendicular bone mineral content', *Medicine and Science in Sports and Exercise*, 16, 223–7.

Woo, S. L-Y., Kuei, S. C., Amiel, D., Gomez, M. A., Hayes, W. C., White, F. C. and Akeson, W. H. (1981) 'The effect of prolonged training on the properties of long bone: a study of Wolff's Law', *Journal of Bone and Joint Surgery*, 65-A (5), 780–7.

Zuckerman, J. D. and Matsen, F. A. (1989) 'Biomechanics of the shoulder', in Nordin, M. and Frankel, V. H. (eds) *Basic Biomechanics of the Skeletal System*, 2nd edn, Philadelphia, PA: Lea & Febiger, pp. 225–47.

10 Mines, meals and movement

A human ecological approach to the interface of 'history and biology'

Holger Schutkowski

Introduction

This chapter deals with activities and roles in past human populations and how these relate to properties of the habitats they live in. Ecological settings and frames significantly shape human habitats, so that their impact on living conditions is at the heart of human ecology. However, human ecology all too often has the connotation of problems related to growing waste pits, overpopulation, resource depletion and parasite load (see e.g. Diesendorf and Hamilton 1997; Nentwig 1995; Southwick 1996). It is true that such problems are human-made, and a reconstruction of historic living conditions has to consider the issues of environmental degradation and its consequences.

However, the ecology of human populations includes more than this blatantly negative aspect. Humans are integral parts of ecosystems, dependent on and tied into structural and functional relations with living organisms and the inanimate environment. Thus the same principles that govern ecological cycles and processes in other organisms apply to human individual life histories and population dynamics, to patterns of spatial dispersion and resource extraction. Even more importantly, humans do not just react to given habitat or ecosystem conditions but are an active agent of change. Apparently, part of the human 'profession' (to paraphrase Eugene Odum's illustrative niche analogy) is to alter landscapes, to bring their influence to bear on species composition and diversity in a habitat, and to steer flows of energy and matter in order to eventually optimise the utilisation of resources on a long-term scale, and thereby safeguard survival in a given habitat.

It is obvious that humans generally seem to be quite successful in doing so, at different points of time and in almost any biome, though to varying degrees. But still more than just the result of this active mediation of ecological constituents, modes and consequences of these actions are equally part of human ecological niches. Unparalleled in scale by any other species, humans make use of non-genetic information that allows them to establish rules and conventions as part of their culture kits, without which the possibilities of environmental structuring could not be transformed and implemented into life-support systems. Thus, in human-shaped ecosystems 'culture' adds to the traditional components of energy, matter and space and provides an enormous storage of socially coded information and experience. Not only is this information quickly retrievable and changeable, but due to the cross-generational mode of exchange it is also available over the long term. Like genetic variation, the information contained in human culture can be passed from generation to generation, and thus has the potential to bridge long time spans. Time, thus, becomes an additional basic ecological component in habitats shaped by humans (Schutkowski 2001).

Hence, an ecologically informed reconstruction of past lifestyle and living conditions should not be limited to identifying humans as permanent destroyers. Rather, it has primarily to comprehend the intentional interference with flows of matter and energy as a decisive feature, without which the evolution and expansion of humans cannot be explained. The diversity of culture kits is closely related to and can, at least partly, be explained by the plasticity of human ecological niches.

These interrelations appear to be readily accessible through data from living populations. One way of retrieving the equivalent information for historic time-frames is to analyse human hard tissues. These form biological archives that allow the extraction of detailed information about the daily life and behaviour of past populations, and they can be deciphered by appropriate analytical methods. This chapter will, thus, draw on trace element and stable isotope data to explore their potential in reconstructing key areas of the biological history of past popula-tions. Three partly interrelated topics will be dealt with briefly by addressing the issues of heavy metal burden and health, social inequality and dietary patterns, and residential mobility.[1] They demonstrate how questions about the past can be illuminated by the application of biological concepts and methods utilising the most direct source material available – the skeletal record.

Historical biomonitoring

In analogy to modern studies in public health, historical biomonitoring relates to the detection of harmful substances that people were exposed to in the past, and their evaluation in terms of health impact. In order to detect and assess hazardous substances in the archaeological record they have to be retrievable from human tissues that regularly survive, such as bones and teeth. Of all trace elements that fall into this category and are toxic even in minor quantities, lead is preferentially stored in the skeleton in the course of a natural detoxification process whereby it is sequestered from the blood stream. It is deposited in the mineral matrix and incorporated into the bioapatite at calcium positions. Other toxic elements, such as cadmium (Cd) and arsenic (As), are also found in hard tissues, even though their typical place of deposition is in inner organs (Cd) and keratinised tissues (As). But since the skeleton makes up a relatively high propor-tion of the total body mass, skeletal concentrations of these elements are still representative of the total body concentration (Grupe 1991). All named elements cause severe damage to the body, the scope ranging from neurotoxic effects to infertility or renal insufficiency (Fergusson 1990).

Studies of heavy metal burden in past human populations have been carried out in a variety of contexts ranging from Romano-British settlements (Waldron 1988) to nineteenth-century North American plantation owners and workers (Aufderheide *et al.* 1985) as well as Barbados slaves (Corruccini *et al.* 1987). Even though the findings significantly helped elucidate the health history of the investigated groups, in none of the studies could the concentration of toxic metals be traced to a known source of origin, however plausible the suggested potential causative effect was. One way to identify definitive sources of toxic contaminants and to relate them to element concentrations in human skeletal remains is to study populations with a known background of hazard exposure. The following example reports on findings from eighteenth-century *silvani* (i.e. those who went into the woods to smelt ores) from the town of Goslar at the north-western rim of the Harz Mountains in Germany. This skeletal sample (N = 89, see Schutkowski *et al.* 2000a for a preliminary report) is among the very rare historic collections with archival evidence of social topography (Klappauf 1996). The miners and smelters of Goslar traditionally inhabited a distinct quarter of the town (the Frankenberg quarter) and were allowed to bury their dead on former monastic premises. It is this context of known occupational risks in a population of known occupation that makes it a unique study in historical biomonitoring.

Since the Bronze Age, central and northern Europe have developed a continuous tradition of mining and smelting, starting with copper and then, from late Roman times, increasingly emphasising silver. During medieval times, in particular, mining of silver became a large-scale enterprise because there was an increasing demand for coin silver. Quite a number of towns, like Goslar, owe their existence and long-term prosperity to the rich mineral veins in the surrounding mountains (Bartels 2000).

Mining and smelting had a visible impact on the natural environment, not only in terms of resource depletion, e.g. by cutting trees for fuel, but also in terms of a deterioration of the environment induced by pollution. Nineteenth-century archival sources illustrate this with reports that due to the deposition of sediment enriched with heavy metals, cattle regularly died from acute lead intoxication in the meadows around the Harz Mountains during the annual flooding of rivers. It was altogether impossible to keep small livestock (Günther 1888). Even today (and therefore often a long time after the activities ceased) traces of intensive mining and smelting can still be detected through so-called geochemical hot spots, anomalies that stick out with extraordinarily high concentrations of heavy metals against the normal background of the surrounding areas (e.g. Fauth *et al.* 1985; Thornton 1988). Such data, however, though telling indicators of past environmental degradation, reflect only the potential health impact of pollution resulting from anthropogenic contamination. The only direct approach to assessing the real extent of heavy metal burden is to analyse the skeletal remains.

The major health problem arising from the processing of ore is the production of waste gases and the emission of dust and smoke filled with the chemical substances from the smelting process. With silver/lead-ores, such as the ones exploited in Goslar, lead in particular, and also to a much lesser extent cadmium, arsenic

Table 10.1 Goslar, eighteenth-century smelters: accumulation of skeletal heavy metal concentrations with age (means and SD, concentrations in µg/g)

Age	0–15	16–30	31–40	41–50	51–60	61+
Lead	97.6 ± 66.8	70.2 ± 31.4	75.5 ± 28.7	99.9 ± 39.2	120.7 ± 80.2	157.7 ± 59.0
Cadmium	2.14 ± 1.0	1.51 ± 0.5	1.63 ± 0.9	1.56 ± 0.9	1.56 ± 0.8	2.29 ± 1.2
Arsenic	2.20 ± 1.5	1.07 ± 0.4	1.42 ± 1.0	1.65 ± 0.4	1.51 ± 0.7	1.99 ± 1.3

Table 10.2 Goslar, eighteenth-century smelters: skeletal heavy metal concentrations in adult individuals (means and SD, concentrations in µg/g)

	Total	Males	Females
Lead	107.8 ± 62.7	106.8 ± 59.8	108.9 ± 69.9
Cadmium	1.71 ± 0.9	1.51 ± 0.9	1.76 ± 0.8
Arsenic	1.60 ± 0.9	1.31 ± 0.6	1.75 ± 1.0

and antimony, have to be considered as airborne toxic substances. Their uptake occurs via both the respiratory tract and food ingestion, with subsequent absorption and transportation to the respective preferential storage tissues.

This leads to characteristic findings for the Goslar *silvani*. Bone lead concentrations show a clear and continuous increase in the mean values with increasing individual age. Since lead is known to accumulate in the skeleton, this corresponds to a continuous lifetime exposure, a finding that corresponds to modern epidemiological data from metalworkers (e.g. Bergdahl *et al.* 1998; Schütz *et al.* 1987). The relatively high values for children and adolescents are most probably due to a higher incorporation rate as a result of increased metabolic activity of the growing skeleton. Cadmium and arsenic also follow the trend of accumulative increase with age, yet to a much lesser extent (Table 10.1).

With regard to gender, there is a conspicuous lack of significant differences in heavy element values between adult males and females. Especially for lead, the values are almost identical to the overall sample mean (Table 10.2). This indicates a similar or at least an equally severe exposure to harmful substances. A straightforward explanation would be that the whole family was engaged in ore smelting and was subject to equal contamination, even though men and women might have performed different tasks. Records of eighteenth-century salt-works indicating that labour was organised on the family level provide an immediate analogue (Witthöft 1996). By this time, however, small-scale ore processing activities in the Harz area had already been replaced by centralised smelting plants (Linke 2000), implying that women were no longer involved in tasks immediately connected with ore smelting. Data from modern environmental and epidemiological studies carried out in the vicinity of metal refinery plants (Fergusson 1990) suggest an alternative explanation, whereby household and street dust highly enriched in heavy metals can be made responsible for contamination contracted indoors. Consequently, even without their being directly involved in the processing of

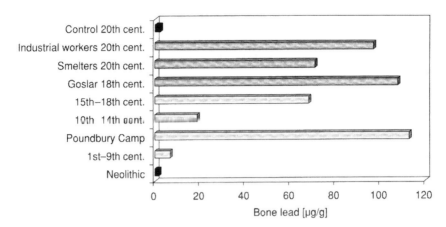

Figure 10.1 Bone lead concentrations on a comparative time scale. There is a general
increase from prehistory into modern times, starting with adaptive non-toxic
values in pre-metal periods (Neolithic). Bone lead levels of eighteenth-century
smelters from a mining region are in the same order of magnitude as values
reported for workers in lead factories and highly industrialized areas today.
Control data are from individuals in areas without heavy metal exposition.
The high figure for the Romano-British site of Poundbury Camp is believed
to be mainly associated with lead from food additives. Modified from
Schutkowski *et al.* 2000a.

metals there was a high and real risk of women being subjected to the same
amount of contamination as the smelters themselves. Contamination of foodstuffs due to the fallout from general waste gas added to it.

Put into the context of a time trend (Figure 10.1) the Goslar *silvani* show lead
levels directly comparable to data derived from modern lead smelting workers
and inhabitants of highly industrialized areas. A general increase of skeletal lead
concentrations through historical time can be observed. The threshold value
of adaptive and potentially non-toxic concentrations, around 1–3 μg/g or even
less (e.g. Budd *et al.* 1998), was generally exceeded during high medieval times,
facilitated by an increase of trade, urbanisation and, later, industrialisation.

Since the skeleton has a relatively long turnover time it can thus serve as the
ideal monitor organ to assess individual long-term accumulative burden and exposure, while blood lead levels, used in clinical investigations, only represent shorter
terms (e.g. Bergdahl *et al.* 1998). Therefore, if recent and increased historic skeletal
concentrations are referred to the empirically known potentially non-toxic values,
it is possible to evaluate those levels of exposure to which humans have been able
to adapt. Thus, only reference data from pre-metal and/or pre-industrial populations allow a true evaluation of adaptations possible to tolerable concentrations of
toxic heavy metals. Both now and in the future, such correlations will be relevant
for pre-emptive measures in public health, e.g. the definition of limit concentrations.
The biomonitoring of past populations thus contributes to an essential aspect of
human ecology that interrelates environmental change and health in past societies.

Residential mobility

Humans may change their place of residence for various reasons, be it in the course of population expansion triggered by a shortage of local food resources, due to climatic change and the need to settle elsewhere, as a result of long distance political or marriage obligations, or for vocational reasons and work opportunities. In doing so, people arrive in a new habitat where the geochemical properties of the bedrock and soil may be different from the place of their departure. The biogeochemistry of Strontium (Sr) allows such migrations to be traced by comparing $^{87}Sr/^{86}Sr$ isotope ratios of mineralised hard tissues.

This issue has been addressed from various angles, for example with regard to the reconstruction of life histories (Sealy *et al.* 1995), cultural shifts (Grupe *et al.* 1997), population movement driven by ecological/environmental change (Price *et al.* 1994), social and ethnic diversification (e.g. Price *et al.* 2000; Schweissing and Grupe 2000) or forensic applications (Beard and Johnson 2000; Schutkowski *et al.* 2001). One immediate reason for a change of domicile, work opportunities, is well known today and presumably was no less important in historic times, though vocational mobility has not been demonstrated yet in archaeological populations. With increasing division of labour and crafts specialisation, however, people with sought-after skills were likely to have been recruited to places of flourishing industries. Thus, within the context of historical mining communities the question arises whether those areas were able to establish a tradition of local trade and craftsmanship or whether they required specialists that had to be searched for from a limited number of trained and able persons throughout the country. The case of medieval *montani*, mining families, from the town of Sulzburg, Black Forest (Steuer 1999), will serve as an example to illustrate the analytical assessment of this major sociocultural factor in connection with the revitalised silver mining business during the Middle Ages.

The general design of such a study makes use of the fact that mineralised tissues of the body possess different material properties. While bone is an organ with a continuous metabolic turnover throughout life, dental enamel is a cell-free tissue and thus not subject to change in its elemental composition apart from minor diffusion at the surface or the enamel-dentin junction. The enamel core, therefore, remains unchanged and reflects the intake of chemical elements during mineralisation in the formation phase. Since food is the major source of elemental supply, the chemical composition of the enamel is determined by the chemical makeup of the food supply. Foodstuffs, in turn, by and large reflect the elemental composition of the soil they were grown on and thus the geology of underlying sediments, groundwater and bedrock. These differences translate into typical strontium isotopic ratios of the local geology. Unlike stable isotopes of lighter elements, strontium isotopes are not subject to fractionation in the course of the food chain or metabolic action due to their higher atomic weight (Graustein 1989). In dental enamel, therefore, the ratio of strontium isotopes in the food and thus the respective isotopic composition of the local geology are preserved. Enamel is becoming an elemental archive of early childhood. Bone, on the other

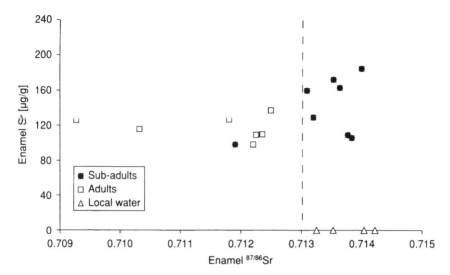

Figure 10.2 Evidence of vocational residential mobility in medieval Black Forest miners. All adult individuals significantly differ in their enamel strontium isotope values from the local water signature indicating that they migrated to the area during their lifetime. Almost all children, however, show signatures identical with local values and are thus isotopically indigenous. Dashed line: suggested cut-off value.

hand, does respond to a change in elemental supply with an alteration of its composition during lifetime, because it is constantly being remodelled. Thus, a change of domicile after dental crown formation will eventually be reflected by a strontium isotopic ratio of the skeleton that is different from that of the enamel, which has still retained the isotopic composition of the person's place of origin.

A scatter plot of ^{87}Sr/^{86}Sr ratios against Sr contents of enamel from the Black Forest miners (Figure 10.2) reveals that the majority of juveniles form a cluster, whereas the adult individuals form two smaller groups separated from the children. Also, the juveniles' values fall within the range of local waters with an average isotopy of 0.7138 (Table 10.3), whereas all adults, an adolescent and one child do not (Schutkowski *et al.* 2000b).[2] During the formation phase of the first molar

Table 10.3 ^{87}Sr/^{86}Sr ratios for medieval miners of the Black Forest, Germany

	Mean	*Standard deviation*
Adults	0.7118	0.0013
Sub-adults*	0.7137	0.0007
Local water	0.7138	0.0004

Notes: *Individual within the isotopic range of adults omitted here: see Figure 10.2.
While sub-adults can be identified as isotopically local, adults migrated into the location from an area with a different isotopic signature.

all but one child lived at Sulzburg, while their (presumed) parents spent their early childhood at other places, because their isotopic ratios differ from those of the Sulzburg area. Almost half of the analysed individuals are therefore likely to have changed their residence during their lifetime and moved to the Black Forest from regions with significantly lower strontium isotopic ratios, e.g. the Rhine Valley. Since most of the children were already born locally, it can be suggested that a new mining community was established after recruited specialists settled in Sulzburg during the twelfth century. This is the first evidence of professional residential mobility in the past established through Sr isotope analyses and it provides evidence that the buoyancy of local crafts and industries was supported and safeguarded by recruitment of non-local specialists.

Isotope ratios are thus a powerful tool to detect culturally induced change in human dispersion patterns. Even though change of residence in past human communities is restricted to localised events in the first place, such studies will eventually lead to the reconstruction of historical migration on a broader scale. Since they monitor movement of people they also indirectly trace gene flow, and are thus an ideal complementary approach to population genetics using ancient DNA (e.g. Bramanti *et al.* 2000).

Dietary behaviour

To achieve long-term settlement and resource use in a given habitat, human populations transform existing natural conditions into lasting means of securing survival and food procurement. Translated into a human ecological context this means participation in as well as manipulation of natural flows of matter and energy by purposeful intervention of their course. It constitutes humans as part of food webs within but often beyond their habitats, yet with a largely cultural mediation of their nutritional behaviour and dietary patterns that are typically orientated towards the appropriation of ecological properties in the respective natural units. It implies that not only the material side of nutrition but also its cultural integration, i.e. its production, availability and meaning are part of these biocultural interrelations. Thus, nutrition is just the visible result of a natural and societal framework that makes up the conditions of food acquisition, i.e. subsistence.

Thus, it is appropriate not to interpret human subsistence in the light of environmental properties alone, but to relate it to sociopolitical conditions and their effects on modes of production. On a cross-cultural scale, such a general correlation can be established, and it forms a basis for classifying human communities (e.g. Johnson and Earle 1987). Evidently, plasticity or rigidity of social organisation either allows far-reaching and flexible or only minimal reactions or adjustments to changing resource circumstances. It can thus be assumed that within a society food production and social differentiation are mutually dependent in a way that social status, among other things, is being defined and stabilised by control over natural resources and means of production. Control, however, at the same time implies the condition of differential access and use of (food) resources. The question remains, however, as to whether this would just lead to a diversified refinement of consumed foodstuffs (see e.g. Bourdieu 1997; Elias 1997) with an overall sufficient supply of the population, or whether differential access to resources would in fact find its expression in measurable variance underlying the consumption of main dietary components by groups of a population.

Placing these interrelations into a human ecological context has two aspects, one concerning energy and matter, the other relating to sociocultural issues. According to ecological fundamentals, plant production pays off energetically, as it achieves a better yield per invested unit energy, whereas the production of an equivalent amount of animal-derived food products is connected with considerably higher energy input and energy consumption. Hence, if resource control is socially variable it may be expected that lower social standing correlates with less energy costs involved in food production, i.e. with an emphasis on plant food based on shorter food chains. A higher social status would allow food production from a longer food chain, resulting in animal husbandry and an improved supply of animal-derived foodstuffs. With that, in turn, the actual determinant for at least parts of flows of energy and matter in a given, human-shaped ecosystem, can be identified, which will be briefly explored for early medieval populations from southern Germany.

In principle, for those historical periods where written documents are available, the reconstruction of such conditions is possible from archival sources and allows the connection of sociopolitical, natural and subsistence factors (e.g. Herrmann 1997). Where written evidence is lacking, or provides only a short time window, this is obviously more difficult, and one has to rely on methods of archaeological sciences aiming at an analysis of material remnants of various sources (see Brothwell and Pollard 2001 for the latest overview). Among those, again, element and isotope analyses (cf. Ambrose and Katzenberg 2000) encompass the analysis of dietary patterns directly from the primary substrate and facilitate an integration of past human populations into the food webs of their habitat.

Dietary reconstruction from trace element levels in the inorganic fraction of bone makes use of the fact that elemental supply from food intake is basically reflected in the mineral. Certain diet-indicating elements, such as strontium and barium, are structurally incorporated into the mineral lattice where they occupy calcium positions. However, it is the total mineral content of the diet which is essential for the transfer of a trace element from diet to bone. This process is dominated by the abundance of calcium and since it is one of the constituting elements of the bone mineral, it will always be preferentially built into the matrix in favour of any other element. This so-called biopurification of calcium (see Elias *et al.* 1982; Burton and Wright 1995) takes place on different trophic levels and between different positions in the food chain. Thus, the abundance of a trace element decreases in the course of this biopurification and the relative trace element concentrations become lower. Plant food is higher in strontium or barium than more biopurified food items such as meat or milk products.

Element intake from dietary supply is also reflected in the organic part of bone, which mainly consists of collagen, and can be detected through the analysis of

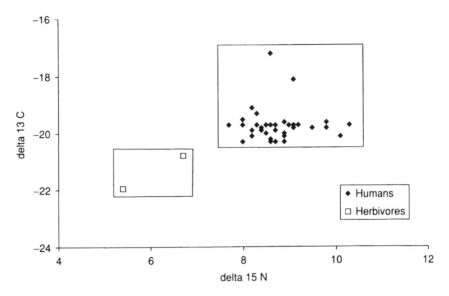

Figure 10.3 Stable isotope ratios for humans and animals at early medieval Weingarten. The trophic level difference between omnivorous humans and herbivores is within the expected range of *c.* 1‰ for carbon and *c.* 3‰ for nitrogen.

stable isotope ratios of carbon and nitrogen; $\delta^{13}C$ values allow the differentiation of terrestrial and marine diets as well as the identification of separate food components, whereas $\delta^{15}N$ ratios can also be used to identify the trophic level of an organism in the food web of its habitat (e.g. Ambrose 1993; Richards and Hedges 1999). When these elements are metabolised, fractionation occurs, which entails a continuous discrimination against the heavier of the stable isotopes. The amount of fractionation varies between the elements, but follows a general consistent pattern of isotopic spacing between different body tissues and trophic positions, which can be used to reconstruct past dietary behaviour.

In the early medieval population of Weingarten (Roth and Theune 1995) this was used to establish very basic food chain relationships between humans and their livestock. It was already known from a previous study (Schutkowski and Herrmann 1996) that people in Weingarten relied on a mixed economy based on the production of various plant and animal-derived foodstuffs. The resulting multi-component omnivorous diet should therefore clearly separate them from herbivore animals. Figure 10.3 shows that animals of known trophic position in the food web differ from the human sample in their isotopic ratios of carbon and nitrogen in the expected amount that denotes isotopic spacing between trophic levels.

The available evidence from trace element analyses at Weingarten identifies a prevalence of dietary components relatively low in mineral/calcium, i.e. a diet presumably largely relying on cereal grains and other non-leafy vegetables with only smaller amounts of milk, milk products and legumes. Meat, being low in mineral, can usually not be detected with its relative contribution to multi-component diets, but the isotopic data suggest that there was at least sufficient supply of animal protein to account for trophic level spacing.

These basic dietary conditions apply to the whole community and are in line with assumptions derived from the ecological potential of the location of Weingarten (N = 143, cf. Schutkowski and Herrmann 1996). The next step is to ask whether sociocultural elements of a community that within a human eco-logical framework are part of the constituting biocultural context can be expected to affect or even govern differential access to food resources.

Early medieval graveyards are known to contain grave goods which can be interpreted to distinguish sociocultural sub-units within a population, comprising groups of individuals sharing similar conditions of wealth. Historical and archaeo-logical interpretations suggest (e.g. Steuer 1989) that early medieval society, unlike the strictly hierarchical social structure of the later Middle Ages, was open and permeable to a high degree. Differences in wealth and status were the result of personally awarded privileges and led to the formation of family groups (*familiae* in late antique tradition) of varying social ranks. Material display of privileges, e.g. in terms of weapons or jewellery, accompanied social differentiation. But since the privileges were acquired and property basically not heritable, there was no necessity to keep status indicating items, except perhaps for emotional com-memoration. Instead, the societal framework fostered the idea of using them as an adornment in the funeral context in order to demonstrate and signify the status of the respective person to everyone in the community. Thus, the burial reflects the actual social status at the time of death in an obvious display of the insignia of rank.

As already pointed out, social differentiation and food production are likely to be interdependent through issues of control over natural resources and the means of production, and thereby responsible for differential access to resources within a population, especially those items that are highly esteemed, such as animal protein. If groups with common dietary behaviours can be identified within skeletal samples, it should be possible to demonstrate whether, and to what extent these dietary groups match social groups.

The Weingarten children and juveniles will be taken as an example here (Figure 10.4; see Schutkowski *et al.* 1999 for further details). Cluster analysis of standardized concentrations of diet-indicating trace elements produced group-ings that can be interpreted as sub-samples reflecting dietary differences. A comparison of these groups and the status affiliation of their members as indic-ated by grave goods shows a significant (p < 0.001) pattern of correspondence. Almost 77 per cent of those individuals who according to grave inclusions would have been members of lower social ranks fall into a cluster that is characterised by a diet likely to contain higher amounts of vegetable items. About 75 per cent of the higher-ranking children and juveniles belong to the cluster that reflects an increased consumption of dietary components rich in calcium, with higher amounts of animal protein. This reveals a general pattern of association between

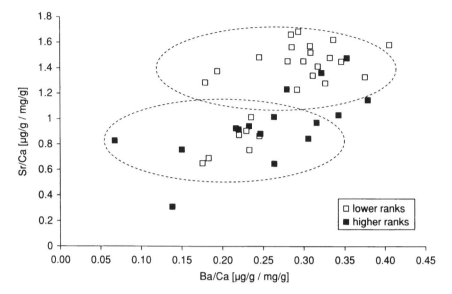

Figure 10.4 Social variation in dietary patterns at Weingarten. The dashed ovals denote dietary groups based on cluster analysis of diet-indicating trace elements of the sub-adult individuals. Individuals of higher social status are significantly associated with element values that reflect better access to dietary protein than those of lower social status.

dietary behaviour and affiliation to a social group. A position of higher rank is generally associated with greater access to high-quality food items. Similar findings apply to the adult individuals, but the contrast is not as pronounced as among the sub-adults.

The pattern of high congruence is significant for two reasons. The allocation of ranks in the early medieval society encompassed all members of the community, so that children as well as adults clearly indicate the social position of the *familia* they belonged to, illustrated by the wealth and composition of their grave equipment. Thus, patterns of congruence between differential nutritional behaviour and social status provide evidence for differences in living conditions between families resulting from differences in rank. Children born into a better-off family also profited from improved living conditions, whereas children in lower ranking families were in a less privileged situation. It can be reasonably assumed that this had bearings on dietary habits and opportunities that were passed on to children in the course of social learning and transmission within the family.

Second, the findings from the non-adult and adult samples corroborate each other, because dietary patterns of children linked to status variation reflect the differential resource opportunities that families of varying ranks had at their disposal. Adults had to face the possibility of changing rank during their lifetime (Weidemann 1982), but while upward social mobility may not necessarily have resulted in personal dietary change in adults, downward mobility almost

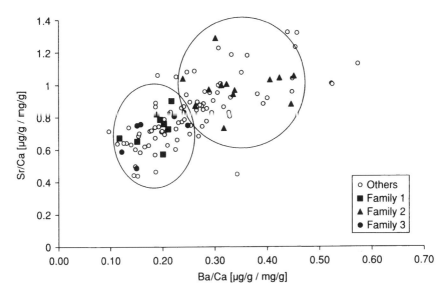

Figure 10.5 Differences in dietary habits are reflected in these presumed families at early medieval Wenigumstadt. Graves denoted 'other' that are adjacent to the family individuals in the scatterplot were checked for their location on the graveyard and confirmed that the clustering of family groups does not simply reflect local soil element contents.

certainly led to a change in living conditions and therefore most likely also to dietary change. In children, however, rank-related nutrition should and does appear in a highly significant way on account of differential resource availability connected with rank-related variation in living conditions, because family differentials in living conditions had an impact on nutritional opportunities and quality for children.

This is supported by further evidence from the contemporary site of Wenigumstadt (Stauch 2000). A conspicuous clustering of three groups of burials on the graveyard, each showing characteristic features of grave architecture and the combination of grave goods, led to an interpretation favouring family burials. Trace element data show (Figure 10.5) that these groups can be distinguished in terms of elemental distribution and thus by their nutritional habits. Accordingly, Family 2 would be characterised by a diet slightly lower in dietary calcium, i.e. with a higher grain component, than Families 1 and 3, whose members are likely to have had better access to animal-derived protein. Even though it might not come as a surprise that families have their typical cuisine, these findings suggest that for the first time the family as *the* constituting unit of the early medieval society can be identified in terms of dietary differences and apparently also differential access to high quality food. This, in turn, corresponds to unequal opportunities of resource utilisation as a result of social rank differentials on the family level in the early medieval society.

Conclusion

Bone chemical analyses prove to be powerful analytical tools that allow the reconstruction of daily activities and life history events from their effects on the skeleton. Furthermore, the data generated can be interpreted within an epistemological framework reaching beyond the short term, and allowing underlying principles of human/environment interaction to be traced. By placing humans in an ecological context, habitat, diet and health can be identified as three major interrelated areas that strongly influence, if not determine, human biological well-being, no less in the past than today. The ecological properties in a given location provide both natural and human-shaped templates that facilitate options of survival, maintenance and expansion available to human populations. While the analytical methods applied here are able to detect biological outcomes of and reactions to different natural and cultural environments in the first place, they also allow tapping the human skeleton as a micro-historic source. It is this decisive step that also makes biological repercussions of cultural mediation accessible.

Heavy metal concentrations in the skeletons of eighteenth-century miners and smelters prove quite suitable as a monitor of serious ongoing biohazards, but they also show that nobody is spared from the consequences of human-induced environmental degradation. The study of local conditions thus reveals their regional and, eventually, also their worldwide (Weiss *et al.* 1999) ecological impact. The preceding step of finding and moving into a habitat of choice is fundamentally connected with the potential and options it provides, and the motivations for individual or population movement are manifold. The example of medieval miners presented here combines the availability of natural resources and the entirely culturally governed aspect of vocational opportunities. It is thus not only mobility itself that can be traced but, moreover, individual and collective decisions that obviously led to the change of domicile. Finally, at the very fundamental level of subsistence options and dietary behaviour energy and matter become inseparable from space, time and culture. The variation in social standing visible in early medieval communities plausibly translates into differential access to high quality food items, which depicts congruent differences in the extraction and utilisation of available energy and resources. Culturally constructed modes of social organisation are thus also reflected in modes of appropriating the ecological potentials of a given habitat.

An approach that applies a conceptual biological framework and operates close to the bone in the attempt to elucidate major themes in human lifestyle and living conditions in the past thus builds a bridge between history and biology.

Acknowledgements

Parts of this study were made possible through funding from the German Research Council and the Volkswagen Foundation, and were carried out at the Department of Historical Anthropology and Human Ecology, University of Göttingen.

Notes

1 For details of analytical methods and technical aspects see references in the text.
2 Issues relating to diagenetic effects are beyond the scope of this chapter. The study suggest, however, that instead of bone, which is likely to suffer from diagenetic change of its isotopic composition, local water serves as a highly suitable reference, instead, against which enamel isotope ratios should be checked.

References

Ambrose, S. H. (1993) 'Isotopic analysis of paleodiets: methodological and interpretive considerations', in Sandford, M. K. (ed.) *Investigations of Ancient Human Tissue*, Langhorne, PA: Gordon & Breach, pp. 59–130.

Ambrose, S. H. and Katzenberg, M. A. (eds) (2000) *Biogeochemical Approaches to Paleodietary Analysis*, New York: Kluwer Academic/Plenum.

Aufderheide, A. C., Angel, J. L., Kelley, J. O., Outlaw, A. C., Outlaw, M. A., Rapp, G. and Wittmers, L. E. (1985) 'Lead in bone III: prediction of social correlates from skeletal lead content in four colonial American populations (Catoctin Furnace, College Landing, Governor's Land, and Irene Mound)', *American Journal of Physical Anthropology*, 66, 353–61.

Bartels, C. (2000) 'Bergbau – ein Überblick' in *Auf den Spuren einer frühen Industrielandschaft: Naturraum, Mensch, Umwelt im Harz*, Arbeitshefte zur Denkmalpflege in Niedersachsen, 21, 106–11.

Beard, B. L. and Johnson, C. M. (2000) 'Strontium isotope composition of skeletal material can determine the birthplace and geographic mobility of humans and animals', *Journal of Forensic Science*, 45, 1049–61.

Bergdahl, I. A., Strömberg, U., Gerhardsson, L., Schütz, A., Chettle, D. R. and Skerfving, S. (1998) 'Lead concentrations in tibial and calcaneal bone in relation to the history of occupational lead exposure', *Scandinavian Journal of Environmental Health*, 24, 38–45.

Bourdieu, P. (1997) *Die feinen Unterschiede: Kritik der gesellschaftlichen Urteilskraft*, Frankfurt am Main: Suhrkamp.

Bramanti, B., Schultes, T., Hummel, S. and Herrmann, B. (2000) 'Genetic characterisation of an historical human society by means of a DNA analysis of autosomal STRs', in Susanne, C. and Bodzsár, É. B. (eds) *Human Population Genetics in Europe*, Budapest: Eötvös University Press, pp. 147–63.

Brothwell, D. R. and Pollard, A. M. (eds) (2001) *Handbook of Archaeological Sciences*, New York and Chichester: Wiley.

Budd, P., Montgomery, J., Cox, A., Krause, P. and Barreiro, B. (1998) 'The distribution of lead within ancient and modern human teeth: implications for long-term and historical exposure monitoring', *The Science of the Total Environment*, 200, 121–36.

Burton, J. H. and Wright, L. E. (1995) 'Nonlinearity in the relationship between bone Sr/Ca and diet: paleodietary implications', *American Journal of Physical Anthropology*, 96, 273–82.

Corruccini, R. S., Aufderheide, A. C., Handler, J. S. and Wittmers, L. E. (1987) 'Patterning of skeletal lead content in Barbados slaves', *Archaeometry*, 29, 233–9.

Diesendorf, M. and Hamilton, C. (eds) (1997) *Human Ecology, Human Economy*, St Leonards, NSW: Allen & Unwin.

Elias, N. (1997) *Über den Prozeß der Zivilisation*, Frankfurt am Main: Suhrkamp.

Elias, R. W., Hirao, Y. and Patterson, C. C. (1982) 'The circumvention of natural biopurification of calcium along nutrient pathways by atmospheric inputs of industrial lead', *Geochimica et Cosmochimica Acta*, 46, 2561–80.

Fauth, H., Hindel, R., Siewers, U. and Zinner, J. (1985) *Geochemischer Atlas der Bundesrepublik Deutschland*, Hannover: Bundesanstalt für Geowissenschaften und Rohstoffe.

Fergusson, J. E. (1990) *The Heavy Elements: chemistry, environmental impact and health effects*, Oxford: Pergamon Press.

Graustein, W. C. (1989) '^{87}Sr/^{86}Sr ratios measure the sources and flow of strontium in terrestrial ecosystems', in Rundel, P. W., Ehleringer, J. R. and Nagy, K. A. (eds) *Stable Isotopes in Ecological Research*, New York: Springer-Verlag, pp. 491–512.

Grupe, G. (1991) 'Anthropogene Schwermetallkonzentrationen in menschlichen Skelettfunden', *Zeitschrift für Umweltchemie und Ökotoxikologie*, 3, 226–9.

Grupe, G., Price, T. D., Schröter, P., Söllner, F., Johnson, C. M. and Beard, B. L. (1997) 'Mobility of Bell Beaker people revealed by strontium isotope ratios of tooth and bone: a study of southern Bavarian skeletal material', *Applied Geochemistry*, 12, 517–25.

Günther, F. (1888) *Der Harz in Geschichts-, Kultur- und Landschaftsbildern*, Hannover: Meyer.

Herrmann, B. (1997) *'Nun blüht es von End' zu End' allüberall' Die Eindeichung des Nieder-Oderbruches 1747–1753*, Münster: Waxmann.

Johnson, A. W. and Earle, T. (1987) *The Evolution of Human Societies: from foraging group to agrarian state*, Stanford, CA: Stanford University Press.

Klappauf, L. (1996) 'Stadtkernarchäologische Untersuchungen in Goslar und die Montanarchäologie des Harzes', *Berichte zur Denkmalpflege Niedersachsen*, 2/96, 53–7.

Linke, F-A. (2000) 'Archaeological survey of monuments of early mining and smelting in the Harz Mountains', in Segers-Glocke, C. and Witthöft, H. (eds) *Aspects of Mining and Smelting in the Upper Harz Mountains*, St Katharinen: Scripta Mercaturae Verlag.

Nentwig, W. (1995) *Humanökologie. Fakten, Argumente, Ausblicke*, Berlin: Springer-Verlag.

Price, T. D., Johnson, C. M., Ezzo, J. A., Ericson, J. and Burton, J. H. (1994) 'Residential mobility in the prehistoric Southwest United States: a preliminary study using strontium isotope analysis', *Journal of Archaeological Science*, 21, 315–30.

Price, T. D., Manzanilla, L. and Middleton, W. D. (2000) 'Immigration and the ancient city of Teotihuacan in Mexico: a study using strontium isotope ratios in human bone and teeth', *Journal of Archaeological Science*, 27, 903–13.

Richards, M. P. and Hedges, R. E. M. (1999) 'A Neolithic revolution? New evidence of diet in the British Neolithic', *Antiquity*, 73, 891–7.

Roth, H. and Theune, C. (1995) *Das frühmittelalterliche Gräberfeld von Weingarten. 1. Katalog der Grabfunde*, Stuttgart: Theiss Verlag.

Schutkowski, H. (2001) 'Human palaeobiology as human ecology', in Brothwell, D. R. and Pollard, A. M. (eds) *Handbook of Archaeological Sciences*, New York and Chichester: Wiley, pp. 219–24.

Schutkowski, H. and Herrmann, B. (1996) 'Geographical variation of subsistence strategies in early mediaeval populations of southwestern Germany', *Journal of Archaeological Sciences*, 23, 823–31.

Schutkowski, H., Wiedemann, F., Bocherens, H., Grupe, G. and Herrmann, B. (1999) 'Diet, status and decomposition at Weingarten: trace element and isotope analyses on early mediaeval skeletal material', *Journal of Archaeological Science*, 26, 675–85.

Schutkowski, H., Fabig, A. and Herrmann, B. (2000a) 'Schwermetallbelastung bei Goslarer Hüttenleuten des 18. Jahrhunderts', in *Auf den Spuren einer frühen Industrielandschaft. Naturraum, Mensch, Umwelt im Harz*, Arbeitshefte zur Denkmalpflege in Niedersachsen, 21, 96–9.

Schutkowski, H., Wormuth, M. and Hansen, B. (2000b) 'Residential mobility of mediaeval Black Forest miners: a strontium isotope study', poster presented at the 12th Congress of the European Anthropological Association, Cambridge, 8–11 September.

Schutkowski, H., Hansen, B. M., Wormuth, M. and Herrmann, B. (2001) 'Signaturen stabiler Strontium-Isotope in menschlichen Hartgeweben – Möglichkeiten für die osteologische Identifikation', in Oehmichen, M. and Geserick, G. (ed.) *Research in Legal Medicine*, 26, 31–40, Lübeck: Schmidt-Römhild.

Schütz, A., Skerfving, S., Mattson, S., Christofferson, J-O. and Alhgren, L. (1987) 'Lead in vertebral bone biopsies from active and retired lead workers', *Archives of Environmental Health*, 42, 340–6.

Schweissing, M. M. and Grupe, G. (2000) 'Local or nonlocal? A research of strontium isotope ratios of teeth and bones on skeletal remains with artificially deformed skulls', *Anthropologischer Anzeiger*, 58, 99–103.

Sealy, J., Armstrong, R. and Schrire, C. (1995) 'Beyond lifetime averages: tracing life histories through isotopic analysis of different calcified tissues from archaeological human skeletons', *Antiquity*, 69, 290–300.

Southwick, C. H. (1996) *Global Ecology in Human Perspective*, New York: Oxford University Press.

Stauch, E. (2000) 'Wenigumstadt – ein Bestattungsplatz der Völkerwanderungszeit und des frühen Mittelalters im nördlichen Odenwaldvorland', *Archäologisches Nachrichtenblatt*, 5, 332–5.

Steuer, H. (1989) 'Archaeology and history: proposals on the social structure of the Merovingian Kingdom', in Randsborg, K. (ed.) *The Birth of Europe: archaeological and social development in the first millennium AD*, Rome: L'Ermadi Bretschneider, pp. 100–23.

Steuer, H. (ed.) (1999) 'Alter Bergbau im Sulzbachtal, Südschwarzwald', *Archäologische Nachrichten aus Baden*, 61/62.

Thornton, I. (1988) 'Soil features and human health', in Grupe, G. and Herrmann, B. (eds) *Trace Elements in Environmental History*, Berlin: Springer-Verlag, pp. 135–44.

Waldron, T. (1988) 'The heavy metal burden in ancient societies', in Grupe, G. and Herrmann, B. (eds) *Trace Elements in Environmental History*, Berlin: Springer-Verlag, pp. 125–33.

Weidemann, M. (1982) *Kulturgeschichte der Merowingerzeit nach den Werken Gregors von Tours*, Mainz: Verlag des RGZM.

Weiss, D., Shotyk, W. and Kempf, O. (1999) 'Archives of atmospheric lead pollution', *Naturwissenschaften*, 86, 262–75.

Witthöft, H. (1996) 'Arbeitsteilung auf der Lüneburger Saline und die Reform des Salzwerks in der Neuzeit (16–19. Jh.)', in Just, R. and Meißner, U. (eds) *Das Leben in der Saline – Arbeiter und Unternehmer*, Quellen zur Kulturgeschichte des Salzes, 3, pp. 302–15.

Index

Page references in *italic* denote figures and tables
Page references followed by the letter n denote material in the notes

Printed and bound by CPI Group (UK) Ltd, Croydon, CR0 4YY

23/10/2024

01778238-0003